# The Essential Workbook for SAT MATH LEVEL 2

- 15 Topic-wise preparation with 245 questions to get ready for SAT Math Level 2 test.
- 15 Minitests with 225 questions with difficulty level ranging from # 40 to # 50 with up-to-date test format.
- Including Answer Keys for Every Question.
- Including Detailed Solutions for Topic-wise Questions / Short Solutions for All Minitests.

Written By

Harim Yoo

Northwestern University(B.A. Mathematics and Economics)

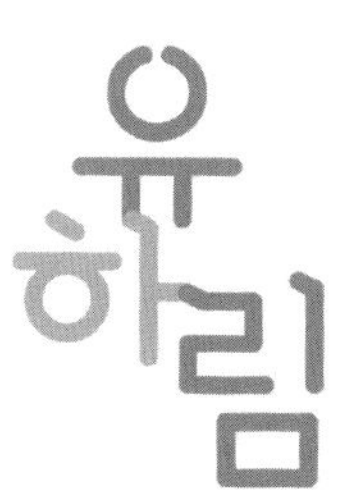


## About the Author

The author, Harim Yoo, graduated from Northwestern University (B.A. in Mathematics) in 2012. Harim also served in the ROK/U.S Army as a KATUSA sergeant in the 2nd infantry division from 2012-2014.
Since his ETS from duty, he found his passion for teaching and educating bright students. He gradually developed his career as a lecturer at Masterprep and has now been recognized as one of the leading lecturers in Apgujeong, Seoul.
Harim is currently dedicating his time to produce mathematicalprep-booksforjuniorhigh/highschools-tudents tosimplifylearningmethodstoeffectivelylearn-thecore concepts and problem-solving strategies, writing on the bulky series called "Essential Math Series", including this book.

# Acknowledgment

This is the first edition of The Essential Workbook for SAT Math Level 2, so I expect to find many places for revision and improvement in each chapter. I have used this book to prepare some of my students for SAT Math Level 2 tests during the summer break, and I reckon it is good enough to be published for the first edition. If you think you have found an error, please let me know through the email birchesinalley@gmail.com, and the edits, along with detailed solutions, will be incorporated in future editions.

Special thanks to my wife, my daughter, and my parents for unconditional support, Mr. Kwon at MasterPrep for unceasing mentorship, and Director Shim as a great role-model and a leader. Finally, I am deeply thankful of God our Lord who led me to this industry ever since 2012 when I graduated from college. I never expected to write any prepbook on highschool mathematics while I was in college. Nonetheless, a series of events in life gave me this priceless opportunity to help students on mathematical journeys during their youth. Thus, I sincerely hope this book turns out to be beneficial for all readers who study with it.

Harim Yoo

# Preface

As all students who prepare for SAT Mathematics Level 2 test would presumably know, the test covers all materials in SAT Mathematics Level 1 test, along with trigonometry and some precalculus materials. Here is how the contents of SAT Math Level 2 test are organized.

- **Algebra and Functions** (35–40%) : Expressions, Equations, Inequalities, Representation and Modeling, Properties of Functions

- **Geometry and Measurement** (35–40%) : 2D Geometry, 3D Geometry, and Trigonometry

- **Data Analysis, Statistics and Probability** (10–15%) : Descriptive Statistics, Graphs and Plots, Regressions, and Probability

- **Number and Operations** (5–10%): Binary Operations, Ratio and Proportion, Complex Numbers, Counting, Elementary Number Theory, Matrices, Sequence and Series, Vectors

I attempted to cover all materials with specific topic titles, most of which would be reiterated in the mini–tests for review purposes. The only part that is missing from this book is the regression part, which I intend to cover in The Essential Guide to SAT Math Level 2. I hope this workbook helps those students in need.

# Contents

# Topic 1

# Linear Functions and Expressions

✓ Slope

✓ Point-Slope Form

✓ Tangency

✓ Perpendicular and Parallel Lines

✓ Slope Sign and Functions

✓ The Meaning and Application of Intersection Points

✓ $\tan(\theta) =$ Slope

## 1.1 Slope

Definition of a slope is

$$\frac{\text{Rise}}{\text{Run}}$$

In other words, it is the ratio of $y$-difference over the $x$-difference. There are three main properties of slopes you have to know.

- Slope of a line containing $(x_1, y_1)$ and $(x_2, y_2)$ equals

$$\frac{y_2 - y_1}{x_2 - x_1}$$

- **Horizontal** line has a slope of 0.
- **Vertical** line has an undefined slope.

---

**MATHEMATICS LEVEL 2 Test - *Continued***

USE THIS SPACE FOR SCRATCH WORK.

1. What is the value of $k$ when a line $2x + ky = 1$ has the slope of 2?

(A) 1
(B) 2
(C) 0
(D) $-2$
(E) $-1$

2. If $x(t) = 2t + 3$ and $y(t) = 4t + 1$, which of the following must be the slope of the line formed by the parametric equations?

(A) $-2$
(B) 3
(C) 2
(D) 1
(E) 4

## 1.2 Point-Slope Form

Similar to the intercept form $y = mx + b$, we also have a point-slope form [1], where

$$y = m(x - h) + k$$

where the graph

- has the slope of $m$
- passes through the point $(h, k)$

There are not many questions to ask the $y$-intercept form, so make sure we get used to this point-slope form.

---

**MATHEMATICS LEVEL 2 Test - *Continued***

USE THIS SPACE FOR SCRATCH WORK.

3. What is the equation of a line with a slope of 4, passing through the point $(6, 2)$?

(A) $y = 4(x - 2) + 6$
(B) $y = 4(x + 2) - 6$
(C) $y = 4(x - 6) + 2$
(D) $y = 4(x - 6) - 2$
(E) $y = \frac{1}{4}(x - 6) + 2$

4. Which of the following explanations about $y = 3x + 6$ is correct?

(A) Its graph is a parabola.
(B) It has the $x$-intercept at $x = 2$.
(C) It has the $y$-intercept at $y = 6$.
(D) The function $y = 3x + 6$ is decreasing.
(E) $y$ increases by 6 as $x$ increases by 1.

[1] Some people call it slope-point form.

## 1.3 Tangency

Had we learned Calculus, we would have used the definition of derivative. However, none of the questions in tangency questions in SAT Math Level II requires the knowledge of Calculus. In fact, there are two types of questions we find during the exam.

- **Geometry** : Tangent line to a circle is **perpendicular** to the line that passes through the point of tangency and the center.

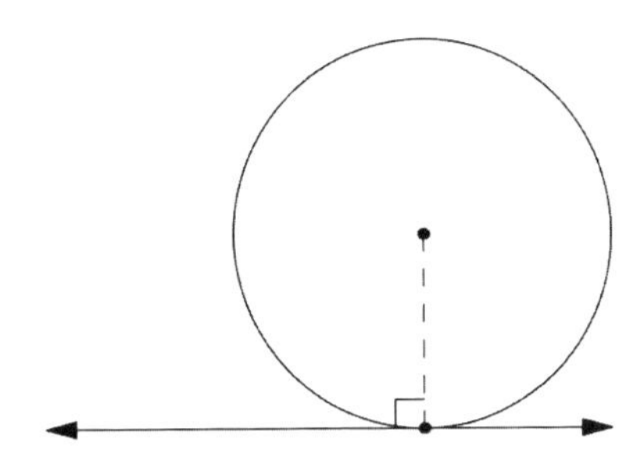

- **System of Equations** : If a quadratic curve and a line are tangent, then there is **one point of intersection**, meaning that when we solve the system of equations, we must get **the discriminant value of** 0.

---

**MATHEMATICS LEVEL 2 Test** - ***Continued***

5. Which of the following lines is tangent to $y = x^2 + 3$?

(A) $y = 4x - 2$
(B) $y = 4x - 1$
(C) $y = 4x$
(D) $y = 4x + 1$
(E) $y = 4x + 2$

USE THIS SPACE FOR SCRATCH WORK.

## 1.4 Perpendicular and Parallel Lines

Given a line and a point off the line, we can always draw two auxiliary lines, i.e., perpendicular or parallel lines. These lines are so special that SAT Math Level 2 question writers cannot help but ask these properties.

- Given $y = mx + b$ and $y = px + q$, if they are **perpendicular**, then

$$m \times p = -1$$

  meaning that the slopes are **negative reciprocals** of one another.

- Given $y = mx + b$ and $y = px + q$, if they are **parallel**, then

$$m = p \text{ and } b \neq q$$

  meaning that they are distinct lines with **equal slopes.**

---

**MATHEMATICS LEVEL 2 Test - *Continued***

USE THIS SPACE FOR SCRATCH WORK.

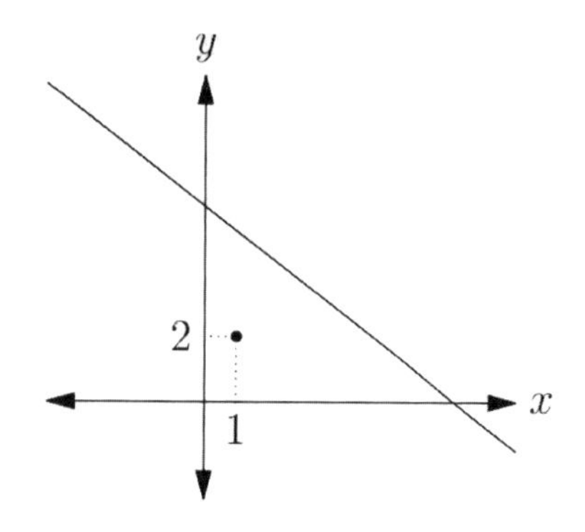


6. Which of the following is the equation of the line that has a point $(1,2)$ on it and is perpendicular to the line $3x + 4y = 24$?

(A) $3x + 4y = 11$
(B) $3x - 4y = -5$
(C) $4x - 3y = 2$
(D) $4x - 3y = -2$
(E) $4x - 3y = 5$

## 1.5 Slope Sign and Function

This usually works for linear functions, but it can be extended to elementary functions we learned in highschool.

- **Slope $> 0$** : the linear function is **increasing** : the graph goes $\nearrow$.
- **Slope $= 0$** : the linear function is **constant** : the graph goes $\longrightarrow$.
- **Slope $< 0$** : the linear function is **decreasing** : the graph goes $\searrow$.

If we extend this to general functions, we get two inequalities such that

- If $a > b$, then $f(a) > f(b)$ : the function $y = f(x)$ is **increasing**.
- If $a > b$, then $f(a) < f(b)$ : the function $y = f(x)$ is **decreasing**.

---

**MATHEMATICS LEVEL 2 Test - *Continued***

USE THIS SPACE FOR SCRATCH WORK.

7. Which of the following functions satisfies that if $a > b$, then $f(a) > f(b)$ for any $a$, $b$ in its domain?

(A) $y = x^2$
(B) $y = |x|$
(C) $y = e^{-x}$
(D) $y = 1/x^2$
(E) $y = x$

8. If $y = f(x)$ is a linear function such that $f(1) \geq f(2)$, $f(3) \leq f(4)$, and $f(0) = 2$, which of the following must be true?

(A) $f(x) = x$
(B) $f(x) = 2x + 1$
(C) $f(x) = 1$
(D) $f(x) = 2$
(E) $f(x) = x + 2$

## 1.6 The Meaning and Application of Intersection Points

Intersection questions generally appear in two forms. The first is a graph form, while the other form is the system of equations form.

When we solve the system of equations, we either **eliminate** or **substitute** one variable into the other. A common solution to the system of equations, if there is any, normally represents the **$x$-coordinate(s)** of the intersection point.

After finding the intersection point, you might have to find other mathematical quantities such as

- the distance between the intersection point and another given point.
- the slope between the intersection point and another given point.

---

**MATHEMATICS LEVEL 2 Test - *Continued***

USE THIS SPACE FOR SCRATCH WORK.

9. What is the distance between the origin and the intersection point between $2x+3y=6$ and $4x+3y=12$?

(A) 0
(B) 1
(C) 2
(D) 3
(E) 4

10. If $f(x)=\sqrt{9-x^2}$, then the slope of a linear function that passes through $(1, f(1))$ and $(-2, f(-2))$ is

(A) 0.144
(B) 0.153
(C) 0.197
(D) 2.941
(E) 3.058

## 1.7 $\tan(\theta) =$ Slope

Given $y = mx + b$, then let's call the angle $\theta$ formed between the $x$-axis and the line in a counterclockwise direction. Then, we get

$$\tan(\theta) = m$$

This is always true because $\frac{\triangle y}{\triangle x}$ comes from a right triangle, and $\triangle y$ measures the opposite side of the angle $\theta$, and $\triangle x$ measures the adjacent side of the angle $\theta$.

---

**MATHEMATICS LEVEL 2 Test - *Continued***

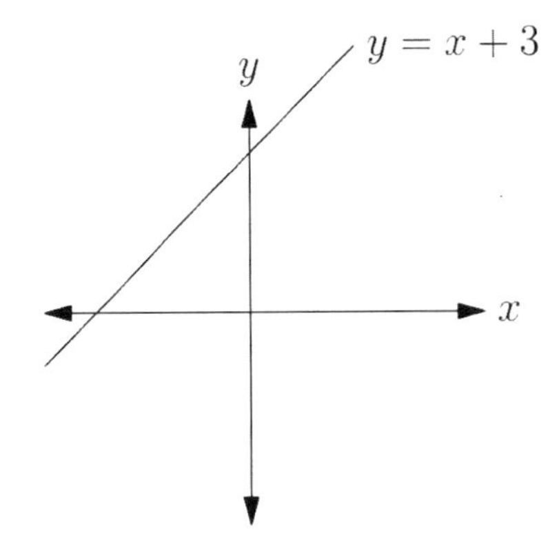


USE THIS SPACE FOR SCRATCH WORK.

11. If $y = x + 3$ and $\theta$ is the acute angle formed by the line and the $x$-axis, what is the measure of $\theta$?

(A) 30° (B) 45° (C) 60° (D) 75° (E) 90°

12. If $\theta$ is an acute angle between $y = x + 3$ and $y = 2x + 3$, what is $\tan\theta$?

(A) $\frac{1}{4}$ (B) $\frac{1}{3}$ (C) $\frac{1}{2}$ (D) $\frac{1}{\sqrt{2}}$ (E) 1

**MATHEMATICS LEVEL 2 Test - *Continued***

USE THIS SPACE FOR SCRATCH WORK.

13. If $y = -2x + 3$ and $\theta$ is the obtuse angle formed by the line and the $x$-axis, what is the value of $\sec(\theta)$?

(A) 3
(B) $-2$
(C) $-\sqrt{5}$
(D) 2
(E) $\sqrt{5}$

14. If $f(x) = mx + b$ where $\tan(\theta) = 3$ for the acute angle $\theta$ formed by the line and the $x$-axis, what is the value of $m$?

(A) $-1$ (B) $\frac{1}{3}$ (C) 3 (D) $-3$ (E) 1

15. If $\theta$ is the angle formed by the line $x = 2$ and the $x$-axis, what is the value of $\tan\theta$?

(A) 1
(B) 2
(C) 3
(D) 4
(E) undefined

## Answer Key to Practice Problems

1. (E)

2. (C)

3. (C)

4. (C)

5. (B)

6. (D)

7. (E)

8. (C)

9. (D)

10. (C)

11. (B)

12. (B)

13. (C)

14. (C)

15. (E)

## Detailed Solution for Practice Problems

**1.**

Given a linear function $f(x) = mx + b$, we say $m$ is the slope of $f(x)$ and $b$ is the $y$-intercept. Let's have a look at what *slope* actually is. The slope of a linear function is defined as the rate of change of $y$ per change of $x$. In other words,

$$\frac{\triangle y}{\triangle x}$$

Given a standard form of a line $Ax + By + C = 0$, it is difficult to see the slope value at a glance, so we will change it into the intercept form, i.e., $y = mx + b$. Let's change the given line equation into $y = f(x)$ form to find out the slope expression.

$$\begin{aligned} 2x + ky &= 1 \\ ky &= -2x + 1 \\ y &= -\frac{2}{k}x + \frac{1}{k} \end{aligned}$$

Hence, $2 = -\frac{2}{k}$ implies that $k = -1$. Therefore, the answer must be **(E)**.

**2.**

Parametric equations are equations with $(x(t), y(t))$ where $x$ and $y$ both depend on the parameter $t$, usually representing for time. The key idea to use when we solve parametric equations is the elimination of the common variable, i.e., $t$. Since $x = 2t + 3$, we may retrieve $2t = x - 3$, so $t = \frac{x}{2} - \frac{3}{2}$. Substituting the expression into $y$, we get

$$\begin{aligned} y &= 4t + 1 \\ &= 4\left(\frac{x}{2} - \frac{3}{2}\right) + 1 \\ &= 2x - 6 + 1 \\ &= 2x - 5 \end{aligned}$$

Hence, the slope must be 2. On the other hand, we could look at the change of $y$ over change of $x$ directly from the parametric equations. Specifically,

$$\frac{\triangle y}{\triangle x} = \frac{4}{2}$$

which is the ratio of the slope of $y(t)$ and $x(t)$. Therefore, the answer must be **(C)**.

**3.**

This is a typical application of a concept called *point-slope* form. Given a slope of 4 and a point $(6,2)$, the line equation must be

$$y = 4(x-6)+2$$

If this is challenging to understand, then think about the fact that the slope between a point $(x,y)$ (on the line) and $(6,2)$ is 4, which can turn into

$$\frac{\triangle y}{\triangle x} = \frac{y-2}{x-6} = 4$$

Hence, $y-2=4(x-6)$, so $y=4x-24+2=4x-22$. The answer must be **(C)**.

**4.**

Let's use the process of elimination. The graph of $y=3x+6$ is a straight line that has a positive slope. (A) is false because its graph is a line. (B) is false because $0=3(-2)+6$ shows that $x=-2$ is the $x$-intercept. (C) is true because $y=3(0)+6=6$ is the $y$-intercept. (D) is false because the function $y=3x+6$ is increasing, illustrated by its positive slope. (E) is false because $y$ increases by 3 as $x$ increases by 1. Thus, the answer is **(C)**.

**5.**

This is a question about intersection between a quadratic curve and a line. Given a parabola $y=x^2+3$ and a line $y=4x+b$ where $b$ is some constant, the line must be tangent to the parabola if $x^2+3=4x+b$ has exactly one real solution. If you see the following phrases such as

- a line *tangent* to a curve
- a quadratic equation has *one real solution*
- a parabola *tangent* to the $x$-axis

then utilize the tool of discriminant, which equals 0, in our case.

$$x^2-4x+(3-b)=0$$

$$\begin{aligned} D &= (-4)^2-4(3-b) \\ &= 16-12+4b \\ &= 0 \end{aligned}$$

Since $b=-1$, the answer must be **(B)**.

**6.**

Perpendicular line has a slope of negative reciprocal to the original line equation. Let's have a look at the original line equation and change it into $y = f(x)$ form.

$$\begin{aligned} 3x + 4y &= 24 \\ 4y &= -3x + 24 \\ y &= -\frac{3}{4}x + 6 \end{aligned}$$

Since the original line equation has the slope value of $-\frac{3}{4}$, its perpendicular line must have the slope value of $\frac{4}{3}$. Also, since the line passes through $(1, 2)$, the line equation must be

$$\begin{aligned} y &= \frac{4}{3}(x-1) + 2 \\ 3y &= 4(x-1) + 6 \\ 3y &= 4x - 4 + 6 \\ 3y - 4x &= 2 \\ 4x - 3y &= -2 \end{aligned}$$

The answer must be **(D)**.

**7.**

- If $a > b$, then $f(a) > f(b)$ = the function $y = f(x)$ is *increasing*.
- If $a > b$, then $f(a) < f(b)$ = the function $y = f(x)$ is *decreasing*.

In our question, the inequality - "if $a > b$, then $f(a) > f(b)$" - implies that the function $y = f(x)$ is increasing for all $x$ in the domain. The only function that satisfies this property is **(E)**.

**8.**

Remember that a linear function is always a straight line that either increases, decreases, or stays unchanged. Given our condition, $f(1) \geq f(2)$ implies that $y = f(x)$ might be decreasing. However, if it is decreasing, we have a problem at $f(3) \leq f(4)$. Hence, $f(1) = f(2)$, as well as $f(3) = f(4)$. Because the function is a linear function, it must be a constant function, meaning that

$$f(x) = f(1) = f(2) = f(3) = f(4)$$

so $f(x) = f(0) = f(1) = \cdots$. The correct answer must be **(D)**.

**9.**

$$\begin{cases} 2x+3y=6 \\ 4x+3y=12 \end{cases} \Longrightarrow \begin{cases} 4x+6y=12 \\ 4x+3y=12 \end{cases}$$

$$\Longrightarrow 3y=0$$

$$\Longrightarrow y=0$$

$$\Longrightarrow x=3$$

The intersection point between $2x+3y=6$ and $4x+3y=12$ can be found by either elimination or substitution. The solution above uses the elimination method to get $(x,y)=(3,0)$. Then, the distance between $(3,0)$ and the origin $(0,0)$ is

$$\begin{aligned} \sqrt{(3-0)^2+(0-0)^2} &= \sqrt{3^2+0^2} \\ &= \sqrt{9} \\ &= 3 \end{aligned}$$

Therefore, the answer must be **(D)**.

**10.**

In order to find the slope, we must find the exact values of $f(1)$ and $f(-2)$.

$$\begin{aligned} f(1) &= \sqrt{9-1^2} \\ &= \sqrt{8} \\ &= 2\sqrt{2} \\ f(-2) &= \sqrt{9-(-2)^2} \\ &= \sqrt{5} \end{aligned}$$

Therefore, the slope between the two points must be equal to

$$\frac{f(1)-f(-2)}{1-(-2)} = \frac{2\sqrt{2}-\sqrt{5}}{3} \approx 0.197$$

The answer must be **(C)**.

**11.**

Using the fact that the slope of a linear function equals the tangent value of the angle formed by the line and the $x$-axis, we get $\tan(\theta)=1$. Out of two special right triangles, $30°-60°-90°$ and $45°-45°-90°$ triangles, $\tan(45°)=1$ whereas $\tan(30°)=\frac{\sqrt{3}}{3}$ and $\tan(60°)=\sqrt{3}$. Or, one could use the calculator to get the value of $\theta$. Hence, $\theta=\tan^{-1}(1)=45°$. Therefore, the answer must be **(B)**.

**12.**

The first possible solution is simple. If we call $\theta_1$ the angle between the $x$-axis and the line $y = x+3$, then $\tan(\theta_1) = 1$. Similarly, if $\theta_2$ is the angle between the $x$-axis and the line $y = 2x+3$, then $\tan(\theta_2) = 2$. If we use a calculator, we may get $\theta_1 = 45°$, then $\theta_2 \approx 63.435°$. Hence, $\theta \approx 18.435°$, so

$$\tan(\theta) = 0.333\cdots = 0.\overline{3} = \frac{1}{3}$$

Therefore, the answer must be **(B)**.

The second possible solution is bit more complicated, but it uses more of trigonometric identities known as

$$\tan(A+B) = \frac{\tan(A)+\tan(B)}{1-\tan(A)\tan(B)}$$
$$\tan(A-B) = \frac{\tan(A)-\tan(B)}{1+\tan(A)\tan(B)}$$

Hence, $\tan(\theta) = \tan(\theta_2 - \theta_1) = \dfrac{\tan(\theta_2)-\tan(\theta_1)}{1+\tan(\theta_2)\tan(\theta_1)} = \dfrac{2-1}{1+2\cdot 1} = \dfrac{1}{3}$, so the answer is **(B)**.

**13.**

First of all, $\tan(\theta) = -2$, so $\theta = \tan^{-1}(-2) \approx -63.435°$.

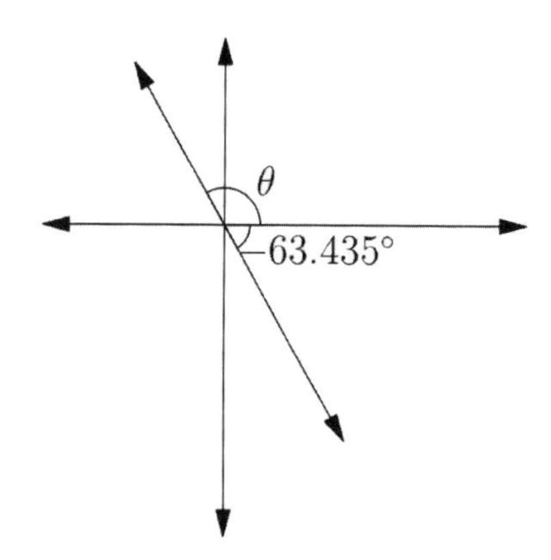


This does not match with the given condition that $\theta$ is obtuse. In fact, as shown in the figure above, the calculator gives back the angle in the 4th quadrant. Since $\tan(A) = \tan(B)$ holds true if $|A-B| = 180°$, we may get another value, which does not directly show up in the calculator, by solving $\theta - (-63.435°) = 180°$, so $\theta = 116.565°$. Hence, $\sec(\theta) = \dfrac{1}{\cos(\theta)} \approx -2.236$. The answer must be **(C)**.

On the other hand, one of the trigonometric identities helps us find $\sec(\theta)$ right away. Specifically, $1+\tan^2(\theta) = \sec^2(\theta)$, so $\sec(\theta) = \pm\sqrt{1+(-2)^2} = \pm\sqrt{5}$. By C.A.S.T rule, shown in Topic 9.1 Trigonometric Ratio, we get $\sec(\theta) = -\sqrt{5}$. Therefore, the answer must be **(C)**.

**14.**

Since $\tan(\theta)$ is the slope of the line equation $y = mx + b$, $\tan(\theta) = 3 = m$. Therefore, the answer must be **(C)**.

**15.**

The vertical line $x = 2$ has no slope. In other words, the slope is undefined. If the slope is undefined, then the associated tangent value must be undefined. Hence, the answer must be **(E)**.

# Topic 2

# Quadratic Functions and Expressions

- ✓ Zeros of Quadratic Equations
- ✓ Properties of Intercepts
- ✓ Discriminant
- ✓ Maximum and Minimum
- ✓ Transformation of Quadratic Graphs
- ✓ Line of Symmetry

## 2.1 Zeros of Quadratic Equations

This is a famous formula named after Vieta. Let's have a look at Vieta's formula. If $r$ and $s$ are solutions to a quadratic equation $ax^2+bx+c=0$, then

- $r+s=-\frac{b}{a}$
- $rs=\frac{c}{a}$

Hard questions in SAT Math Level 2 use this formula, so make sure we actually understand the process.

$$\begin{aligned} a(x-r)(x-s) &= a(x^2-(r+s)x+rs) \\ &= ax^2-a(r+s)x+a(rs) \\ &= ax^2+bx+c \end{aligned}$$

where $b=-a(r+s)$ and $c=a(rs)$, so $-\frac{b}{a}=r+s$ and $\frac{c}{a}=rs$.

---

**MATHEMATICS LEVEL 2 Test - *Continued***

USE THIS SPACE FOR SCRATCH WORK.

1. If $\sqrt{3}$ and $\sqrt{2}$ are two zeros of the equation $x^2+mx+n=0$, then $m^2-2n=$

(A) $\sqrt{3}+\sqrt{2}$
(B) $\sqrt{3}-\sqrt{2}$
(C) 5
(D) 1
(E) $-5$

2. If $1-i$ is a solution to $x^2+mx+n=0$, where $m$, $n$ are integers, what must be the value of $n$?

(A) $1+i$
(B) 2
(C) $2i$
(D) $-2$
(E) $-2i$

## 2.2 Properties of Intercepts

Let's categorize intercepts into two possible cases. The first case to look at is the $x$-intercept.

$x$-intercept questions directly utilize the **discriminants, factorization, or vertex form**. Also, given the $x$-intercepts, we can get the $x$-coordinates of the vertex form. Suppose $r$ and $s$ are $x$-intercepts of $y = ax^2 + bx + c$, then

- The axis of symmetry $x = \dfrac{r+s}{2}$
- The axis of symmetry $x = -\dfrac{b}{2a}$

On the other hand, $y$-intercept is directly derived from the standard form of a quadratic function $y = ax^2 + bx + c$. It is simply $c$.

- $x$-intercept(s) : Set $y = 0$ to find the $x$-intercept(s).
- $y$-intercept : Set $x = 0$ to find the $y$-intercept.

---

**MATHEMATICS LEVEL 2 Test - *Continued***

USE THIS SPACE FOR SCRATCH WORK.

3. If $(t, 0)$ is a point on the curve $y = 2x^2 - x - 3$ and the positive $x$-axis, then $t =$

(A) $-1$ (B) $\dfrac{2}{3}$ (C) 1 (D) $\dfrac{3}{2}$ (E) $-\dfrac{2}{3}$

4. Which of the following must be true about the quadratic function $y = ax^2 + bx + c$ for $a \neq 0$?

(A) There is at least one $x$-intercept.
(B) There is one $y$-intercept.
(C) The graph is symmetric about the $y$-axis.
(D) The vertex is always above the $x$-axis.
(E) The graph is concave up.

## 2.3 Discriminant

Given $y=ax^2+bx+c$, we can have a look at $a$. First off, $a \neq 0$ in order to maintain its quadratic property.

- $a>0$ : the graph is **concave up**, and it has a **minimum** value.
- $a<0$ : the graph is **concave down**, and it has a **maximum** value.

Now, let's have a look at discriminant, which tells us the number of **real solution**[1]. Given $y=ax^2+bx+c$, we define $D=b^2-4ac$ such that

- $D>0$ : the graph intersects the $x$-axis at two distinct points : **2 distinct real solutions**.
- $D=0$ : the graph intersects the $x$-axis at one point : **1 real solution** (in fact, two repeated solutions).
- $D<0$ : the graph does not intersect the $x$-axis : **0 real solution** (but two complex solutions, by the Fundamental Theorem of Algebra).

---

**MATHEMATICS LEVEL 2 Test - *Continued***

USE THIS SPACE FOR SCRATCH WORK.

5. Which of the following functions has three distinct real roots?

(A) $y=(x^2+1)(x^2+3)$
(B) $y=(x^2+1)(x^2-4)$
(C) $y=(x^3-1)(x^2+4)$
(D) $y=(x^2+x-1)(x-1)$
(E) $y=(x^2-1)(x^4-1)$

[1] In SAT Math Level 2, there are synonyms for **solution** - **root**, **$x$-intercept**, and **zero**.

## 2.4 Maximum and Minimum

Quadratic function always has maximum (or minimum) $y$-value. In fact, this is the most common type of qudratic function questions found in SAT Math Level 2. There are two types of questions for maximum and minimum.

- Directly find the maximum (or minimum) value by **completing the square**.
- Find the range of the quadratic function by looking at the **vertex form**.

Given $y = ax^2 + bx + c$, we should be able to transform it into $y = a(x-h)^2 + k$ where

- $x = h$ is the axis of symmetry.
- $y = k$ is either maximum (or minimum) value. If $a > 0$, then it must be minimum. On the other hand, if $a < 0$, then it must be maximum.

Let's have a look at a question about completing the square.

---

**MATHEMATICS LEVEL 2 Test - *Continued***

USE THIS SPACE FOR SCRATCH WORK.

6. The height of an object is given by $h(t) = -5t^2 + v_0 t + h_0$, where $t$ is time in seconds, $v_0$ the initial velocity in meter per second, and $h_0$ the initial height in meters. What is the maximum height of an object when $v_0 = 20$ and $h_0 = 9$?

(A) 9 meters
(B) 20 meters
(C) 29 meters
(D) 30 meters
(E) 34 meters

7. If $y = x^2 + 4x + 3$ for $-3 \leq x \leq 0$, the minimum value of $y$ occurs at $x =$

(A) $-1$ (B) $-2$ (C) $-3$ (D) $-4$ (E) $-5$

## 2.5 Transformation of Quadratic Graphs

This is not just limited to the transformation of quadratic graphs, but shifting transformation frequently appears in the format of quadratic function questions. Remember that

- **Vertex** is the key point for quadratic graphs, i.e., keep focusing on how the vertex is shifted.
- **Translation(or shifting)** is usually the **last**.

How about reflection? Let's have a look at reflection.

- $y = f(x) \longrightarrow y = f(-x)$ : Reflection about **the $y$-axis.**
- $y = f(x) \longrightarrow y = -f(x)$ : Reflection about **the $x$-axis.**
- $y = f(x) \longrightarrow y = -f(-x)$ : Reflection about **the origin**, i.e., reflect it about **the $x$-axis and $y$-axis.**

Dilation can be categorized into two types : vertical stretch(shrink) or horizontal stretch(shrink).

- $y = f(x) \longrightarrow y = f(kx)$ : Horizontal shrink by a factor of $k$, never interfering with the height. For instance, $y = f(2x)$ horizontally shrinks the graph of $y = f(x)$ by a factor of 2, while keeping the height unchanged.
- $y = f(x) \longrightarrow y = kf(x)$ : Vertical stretch by a factor of $k$, never interfering with the width. For instance $y = 2f(x)$ vertically strecthes the graph of $y = f(x)$ by a factor of 2, while keeping the width unchanged.

---

MATHEMATICS LEVEL 2 Test - *Continued*

8. If the graph of $y = 2x^2$ is reflected about the $x$-axis, translated 2 units left and 3 units up, where is the vertex of the resulting graph?

USE THIS SPACE FOR SCRATCH WORK.

(A) $(0,0)$
(B) $(-2,3)$
(C) $(2,3)$
(D) $(3,2)$
(E) $(3,-2)$

**MATHEMATICS LEVEL 2 Test - *Continued***

9. Which of the following functions has the smallest minimum value?

USE THIS SPACE FOR SCRATCH WORK.

(A) $y = (x-1)(x-3)$
(B) $y = -(x-1)(x-3)$
(C) $y = 2(x-2)(x-4)$
(D) $y = 3(x-3)(x-5)$
(E) $y = 4(x-4)(x-6)$

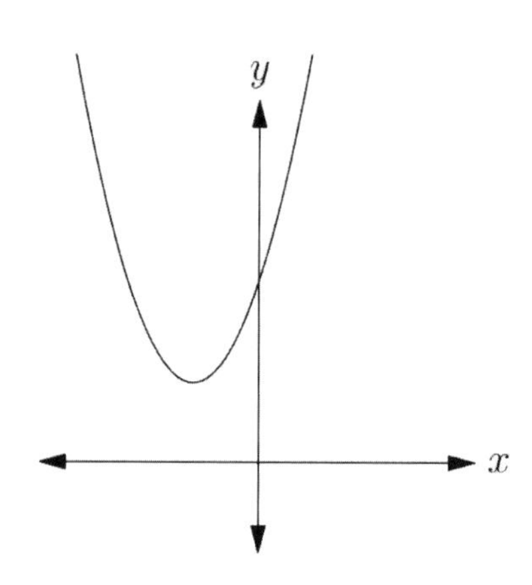


10. If the graph of $y = x^2 + 3x + 4$ is translated right by 2, which of the following must be true?

(A) The range is unchanged.
(B) The vertex is unchanged.
(C) The $y$-intercept is unchanged.
(D) The $x$-intercepts are unchanged.
(E) The axis of symmetry is unchanged.

## 2.6 Line of Symmetry

Given a quadratic function $y = a(x-h)^2 + k$,

- $x = h$ is the line of symmetry.
- $y = k$ is either the minimum (or maximum) value of the function.

Because the line $x = h$ is the line of symmetry, if there are two points $(a, f(a))$ and $(b, f(b))$ where $f(a) = f(b)$, then

$$h = \frac{a+b}{2}$$

For example, if $f(x)$ is a quadratic function such that $f(1) = f(5)$, then the $x$-coordinate of the vertex is $\frac{1+5}{2} = 3$.

---

**MATHEMATICS LEVEL 2 Test - *Continued***

11. If the graph of $f(x) = ax^2 + bx + c$ satisfies $f(4-x) = f(x+8)$, which of the following is the value of $-\frac{b}{2a}$?

(A) 3
(B) 6
(C) 1
(D) −6
(E) −3

12. Which of the following functions satisfies $f(-x) = f(x)$?

(A) $y = \sin(x)$
(B) $y = \tan(x)$
(C) $y = e^x$
(D) $y = \ln(x)$
(E) $y = x^2$

**MATHEMATICS LEVEL 2 Test - *Continued***

USE THIS SPACE FOR SCRATCH WORK.

13. If $f(x) = ax^2 + bx + c$ where the $x$-intercepts are 3 and 7, which of the following must be the value of $\frac{b}{2a}$?

(A) 3 (B) 5 (C) 7 (D) $-5$ (E) 21

14. If $f(x) = x^2 + 24x + 143$, the minimum value of $f(x)$ occurs at $x =$

(A) 11
(B) 12
(C) 13
(D) $-12$
(E) $-11$

15. If $(2, 5)$ satisfies $y = ax^2 + bx + c$ where the line of symmetry is $x = 1$, which of the following values of $x$ must have the $y$-value of 5?

(A) 0
(B) 1
(C) 2
(D) 4
(E) 5

## Answer Key to Practice Problems

1. (C)

2. (B)

3. (D)

4. (B)

5. (D)

6. (C)

7. (B)

8. (B)

9. (E)

10. (A)

11. (B)

12. (E)

13. (D)

14. (D)

15. (A)

## Detailed Solution for Practice Problems

**1.**

First, let's use the fact that two roots are given to us. If $\sqrt{3}$ and $\sqrt{2}$ are two zeros of the equation $x^2+mx+n=0$, then

$$\begin{aligned} x^2+mx+n &= (x-\sqrt{3})(x-\sqrt{2}) \\ &= x^2-(\sqrt{3}+\sqrt{2})x+\sqrt{3}\sqrt{2} \\ &= x^2-(\sqrt{3}+\sqrt{2})x+\sqrt{6} \\ m &= -(\sqrt{3}+\sqrt{2}) \\ n &= \sqrt{6} \end{aligned}$$

Hence, $m^2-2n=(-(\sqrt{3}+\sqrt{2}))^2-2\sqrt{6}=(3+2\sqrt{6}+2)-2\sqrt{6}=5$. On the other hand, by the Vieta's formula,

- $-\frac{m}{1}=\sqrt{3}+\sqrt{2}$
- $\frac{n}{1}=\sqrt{3}\times\sqrt{2}$

Either way, the answer must be **(C)**.

**2.**

Follow this rule if one of the zeros is given. This back-steps the solving mechanism of quadratic equation.

$$\begin{aligned} x &= 1-i \\ x-1 &= -i \\ (x-1)^2 &= (-i)^2 \\ x^2-2x+1 &= -1 \\ x^2-2x+2 &= 0 \end{aligned}$$

Since $x^2-2x+2=x^2+mx+n$, the value of $n$ must be 2. On the other hand, if $1-i$ is a root, then $1+i$ is also the root because the coefficients are all real. Hence,
$(x-(1-i))(x-(1+i))=x^2-(1-i+1+i)x+(1-i)(1+i)=x^2-2x+2$ is retrieved. Hence, the answer is **(B)**.

**3.**

If $(t,0)$ is on the curve $y=2x^2-x+3$, then $2t^2-t+3=0$. In other words, we are looking for $x$-intercepts of the curve $y=2x^2-2x+3$. Solving it by factoring results in $(2t+3)(t-1)=0$. Hence, $t=1$, so the answer is **(C)**.

**4.**

The quadratic function always has one $y$-intercept, no matter what. Think about it this way. If we set $x = 0$, we always get the value of $y$-intercept. No matter what happens to the function, $y = a(0)^2 + b(0) + c$ is always $y = c$. Therefore, the $y$-intercept always exists.

- (A) has counterexample if $y = x^2 + x + 2$.
- (C) has counterexample if $y = (x+1)^2 + 1$.
- (D) has counterexample if $y = (x-1)^2 - 1$.
- (E) has counterexample if $y = -x^2$

As one can see clearly from the bullet-points, all the other answer choices have counterexamples. Hence, the answer is **(B)**.

**5.**

- (A) has no real root because $x^2 + 1 = 0$ and $x^2 + 3$ have no real root. In fact, $x^2 + \bigstar = 0$ has no real root.
- (B) has two real roots because $x^2 - 4 = 0$ has two solutions.
- (C) has one real solution because $x^3 - 1 = 0$ has one real solution $x = 1$.
- (D) has three solutions because $x^2 + x - 1 = 0$ has two solutions since $D = 1^2 - 4(-1)(1) > 0$ and $x - 1 = 0$ has one solution $x = 1$.
- (E) has two solutions because $x^2 - 1 = 0$ has two solutions and $x^4 - 1 = (x^2-1)(x^2+1) = 0$ has two solutions that are repeated from the previous factor.

Therefore, the answer is **(D)**.

**6.**

Let's find the vertex form of $h(t) = -5t^2 + 20t + 9$ to find out the maximum height of an object.

$$\begin{aligned}
-5t^2 + 20t + 9 &= -5(t^2 - 4t) + 9 \\
&= -5(t^2 - 4t + 4 - 4) + 9 \\
&= -5(t^2 - 4t + 4) + (-5)(-4) + 9 \\
&= -5(t^2 - 4t + 4) + 29 \\
&= -5(t-2)^2 + 29
\end{aligned}$$

The maximum height occurs at 29 meters at $t = 2$ (seconds). Hence, the answer is **(C)**.

**7.**

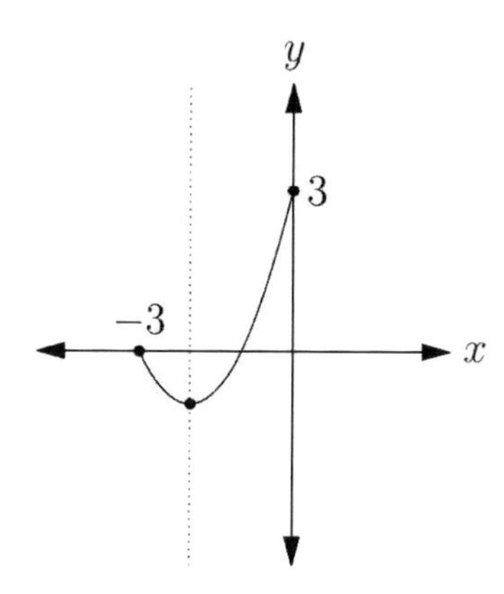


As one can see from the figure above, the graph of $y = x^2 + 4x + 3$ for $-3 \le x \le 0$ reaches its minimum value at the vertex. Let's find the vertex form.

$$\begin{aligned} x^2 + 4x + 3 &= (x^2 + 4x + 4 - 4) + 3 \\ &= (x^2 + 4x + 4) + (-4) + 3 \\ &= (x+2)^2 - 1 \end{aligned}$$

The vertex is $(-2, -1)$, so the minimum value of $y$ occurs at $x = -2$. The answer must be **(B)**.

**8.**

1. Reflect $y = 2x^2$ about the $x$-axis to get $y = -2x^2$.
2. Translate $y = -2x^2$ 2 units left to get $y = -2(x+2)^2$.
3. Translate $y = -2(x+2)^2$ 3 units up to get $y = -2(x+2)^2 + 3$.

The vertex is now at $(-2, 3)$, so the answer must be **(B)**.

**9.**

Let's use the fact that the vertex is located at the midpoint of the $x$-intercepts, if any.

- For (A), the midpoint of the $x$-intercepts is 2.
- For (B), the midpoint of the $x$-intercepts is 2, as well.
- For (C), the midpoint is 3. For (D), the midpoint is 4.
- For (E), the midpoint is 5.

If we substitute all these values into respective quadratic expressions, we get the smallest value for (E). Hence, the answer must be **(E)**.

**10.**

The quadratic function has its lowest(or highest) point at its vertex. In this question, the graph of $y = x^2 + 3x + 4$ is concave up, so it has the lowest point at the vertex. Specifically,

$$\begin{aligned} y &= x^2 + 3x - 4 \\ &= (x^2 + 3x + \frac{9}{4}) - \frac{9}{4} - 4 \\ &= (x + \frac{3}{2})^2 - \frac{25}{4} \end{aligned}$$

where the vertex is at $\left(-\frac{3}{2}, -\frac{25}{4}\right)$. If the vertex is translated either right or left, the original minimum value will not change, meaning that the range will not change. Hence, the answer must be **(A)**.

**11.**

Since $f(4-x) = f(x+8)$, we get

$$\begin{aligned} f(4-x) &= f(x+8) \\ a(4-x)^2 + b(4-x) + c &= a(x+8)^2 + b(x+8) + c \\ a(16 - 8x + x^2) + 4b - bx + c &= a(x^2 + 16x + 64) + bx + 8b + c \\ 16a - 8x + ax^2 + 4b - bx + c &= ax^2 + 16x + 64a + bx + 8b + c \\ 16a - 8x + 4b - bx &= 16x + 64a + bx + 8b \\ 16a + 4b - (8+b)x &= (16+b)x + 64a + 8b \\ 0 &= (24 + 2b)x + (48a + 4b) \end{aligned}$$

Hence, $2b = -24$ and $4b = -48a$, so $b = -12$ and $a = 1$ Therefore, $-\frac{b}{2a} = -\frac{-12}{2(1)} = 6$. The answer must be **(B)**.

That being said, the shortcut for this type of question is that a quadratic function $y = ax^2 + bx + c$ satisfies the following property : if $f(A) = f(B)$, then the midpoint of $A$ and $B$ must be the $x$-coordinates of the vertex. The midpoint of $4 - x$ and $x + 8$ is $\frac{4 - x + x + 8}{2} = 6$, and $-\frac{b}{2a}$ is equal to the $x$-coordinates of the vertex. Hence, the answer is **(B)**.

**12.**

What is the meaning of $f(-x) = f(x)$? It means that $(-x, y)$ and $(x, y)$ are on the graph of the function,implying that the line of symmetry is at $x = 0$. If a function $f(x)$ satisfies $f(-x) = f(x)$ for all values of $x$ in the domain, then the function is called *even*. The only answer choice that satisfies this property is (E). In fact, if a function $y = f(x)$ satisfies $f(-x) = f(x)$, it *evens out* the negative sign. Therefore, the answer is **(E)**.

**13.**

Let's use the shortcut I mentioned in question 11. Notice that

$$-\frac{b}{2a} = x\text{-coordinates of the vertex}$$

Recall that the $x$-coordinates of the vertex is the average of the two $x$-values with the same $y$-values. In this case, we could use it for $x$-intercepts because $x$-intercepts have the same $y$-value of 0. Thus,

$$\begin{aligned}\frac{3+7}{2} &= -\frac{b}{2a}\\ 5 &= -\frac{b}{2a}\\ -5 &= -\left(-\frac{b}{2a}\right)\\ -5 &= \frac{b}{2a}\end{aligned}$$

Therefore, the answer must be **(D)**.

**14.**

Completing the square, we get $f(x) = x^2 + 24x + 143$ as

$$\begin{aligned}f(x) &= x^2 + 24x + 143\\ &= (x^2 + 24x + 144 - 144) + 143\\ &= (x^2 + 24x + 144) + (-144) + 143\\ &= (x+12)^2 - 1\end{aligned}$$

Hence, the minimum value of $y = f(x)$ occurs at $x = -12$. The answer must be **(D)**.

**15.**

Since the line of symmetry is $x = 1$, then the two $x$-values with equal $y$-values will have the average value of 1. In other words, $f(2) = f(a)$ holds for the quadratic function $y = f(x)$ if

$$\begin{aligned}\frac{2+a}{2} &= 1\\ 2+a &= 2\\ a &= 0\end{aligned}$$

The answer must be **(A)**.

# CHECK ON LEARNING #1

## Quadratic Formula

$$x = \frac{-b \pm \sqrt{b^2 - 4ac}}{2a} \text{ for } ax^2 + bx + c = 0$$

Given $ax^2 + bx + c = 0$, the quadratic formula can be deduced by the following set of equations.

$$ax^2 + bx + c = 0$$
$$a(x^2 + \frac{b}{a}x) + c = 0$$
$$a(x^2 + \frac{b}{a}x + \left(\frac{b}{2a}\right)^2) + c = a(\frac{b^2}{4a^2})$$
$$a(x + \frac{b}{2a})^2 + c = \frac{b^2}{4a}$$
$$a(x + \frac{b}{2a})^2 = \frac{b^2}{4a} - c$$
$$(x + \frac{b}{2a})^2 = \frac{b^2 - 4ac}{4a^2}$$
$$x + \frac{b}{2a} = \pm\sqrt{\frac{b^2 - 4ac}{4a^2}}$$
$$x = -\frac{b}{2a} \pm \frac{\sqrt{b^2 - 4ac}}{2a}$$
$$x = \frac{-b \pm \sqrt{b^2 - 4ac}}{2a}$$

# Topic 3

# Exponential Functions and Expressions

✓ Finding $x$-values

✓ Increasing or Decreasing Function

✓ Power Expressions

✓ Inequalities

✓ Exponential Modeling

## 3.1 Finding $x$-values

There are several types of exponential equations in SAT Math Level 2.

- The easiest form is when the bases are equal.
- The intermediate form is when the bases are different.
- The challenging form is when the exponential expression is in quadratic form.

Equalizing the base is the crucial technique to solve exponential equations. However, if this is not possible, we either put log or ln in front of the equation to solve for $x$.

The most challenging form you could find in the exam is when the exponential equation is in the quadratic form.

---

**MATHEMATICS LEVEL 2 Test - *Continued***

USE THIS SPACE FOR SCRATCH WORK.

1. Which of the following is the solution for $3^x = 27^{x-1}$?

(A) 0
(B) 1
(C) 0.67
(D) 1.33
(E) 1.5

2. If $2^{2x} - 2^x - 2 = 0$, which of the following could be the value of $x$?

(A) $-1$
(B) 1
(C) 0
(D) 2
(E) $-2$

**MATHEMATICS LEVEL 2 Test - *Continued***

USE THIS SPACE FOR SCRATCH WORK.

3. What power of 9 is equal to $\frac{1}{27}$?

(A) $-3$ (B) $\frac{3}{2}$ (C) $-\frac{2}{3}$ (D) $-\frac{3}{2}$ (E) $\sqrt{3}$

4. If $f(x) = 10^{2x-1}$ for all real $x$, and if $f(t) = 1{,}000{,}000$, then $t =$

(A) 6 (B) $-\frac{7}{2}$ (C) $\frac{2}{7}$ (D) $\frac{7}{2}$ (E) $-6$

5. Given an exponential expression $y = a \cdot x^k$, $y = 2$ when $x = 4$, or $y = 4$ when $x = 2$. Which of the following is the value of $k$?

(A) $-2$ (B) $\frac{1}{2}$ (C) $-1$ (D) $-\frac{1}{2}$ (E) 2

## 3.2 Increasing or Decreasing Function

Exponential function is either strictly increasing or decreasing function. In fact, this is a type of 1-to-1 function.

- If $a > b$, then $f(a) > f(b)$ : $y = b^x$ where $1 < b$.
- If $a > b$, then $f(a) < f(b)$ : $y = b^x$ where $0 < b < 1$.

In either case, the graph has a horizontal tangent $y = 0$. In other words,

- $\lim_{x \to -\infty} b^x = 0$ where $1 < b$.
- $\lim_{x \to \infty} b^x = 0$ where $0 < b < 1$.

---

**MATHEMATICS LEVEL 2 Test - *Continued***

USE THIS SPACE FOR SCRATCH WORK.

6. Which of the following CANNOT be in the range of $y = 30 + 20e^{-x}$?

(A) 25
(B) 35
(C) 45
(D) 55
(E) 65

7. Which of the following functions satisfies that if $a \neq b$, then $f(a) \neq f(b)$ for all $a$ and $b$ in the domain?

(A) $y = \lfloor x \rfloor$
(B) $y = x^2$
(C) $y = |x|$
(D) $y = e^{-x}$
(E) $y = x^{-2}$

## 3.3 Power Expressions

SAT Math Level 2 frequently asks students to be able to simplify power expression.

- $a^x a^y = a^{x+y}$
- $a^x / a^y = a^{x-y}$
- $a^0 = 1$
- $a^1 = a$
- $a^{-1} = 1/a$

---

**MATHEMATICS LEVEL 2 Test - *Continued***

USE THIS SPACE FOR SCRATCH WORK.

8. If $a$, $b$, and $c$ are nonzero real numbers such that $a^2b^4c^3 = \dfrac{9b^3c^3}{a^{-2}}$, then the value of $b$ equals

(A) $-\frac{1}{3}$ (B) $-9$ (C) 3 (D) 9 (E) $\frac{1}{3}$

9. If $x$, $y$, and $z$ are consecutive integers that add up to 9, what is the value of $a$ if $a = \sqrt[3]{2^x}\sqrt[3]{2^y}\sqrt[3]{2^z}$?

(A) 2
(B) 6
(C) 8
(D) 16
(E) 512

**MATHEMATICS LEVEL 2 Test - *Continued***

10. If $3^x + 3^x = k3^{x+1}$ for $a > 1$ and all real $x$, then $k =$

USE THIS SPACE FOR SCRATCH WORK.

(A) $\frac{3}{2}$ (B) $-2$ (C) 1 (D) 2 (E) $\frac{2}{3}$

11. If $f(x-2) = \frac{1}{4}f(x)$ for all $x$, which of the following functions could be $y = f(x)$?

(A) $f(x) = x$
(B) $f(x) = x^2$
(C) $f(x) = 2^x$
(D) $f(x) = \ln(x)$
(E) $f(x) = |x|$

12. If $m^{2n} = x$, then $m^{8n} =$

(A) $3x$
(B) $4x$
(C) $x$
(D) $x^4$
(E) $x^3$

## 3.4 Inequalities

Inequalities typically use the sign of numbers. For instance, if $ab > 0$, then we either get $(+)(+) > 0$ or $(-)(-) > 0$. Similarly, if $ab < 0$, then we get $(-)(+) < 0$ or $(+)(-) < 0$. Adding or subtracting a constant to the given inequality does not switch the sign, but multiplying by a negative constant does change the sign. For instance, $a > b$ is switched to $-a < -b$.

There are two types of inequalities involving the exponential functions and expressions.

- $(\qquad)^x > 0$ for all $x$. In other words, there is no $x$-intercept.
- if $x > y$, then $b^x > b^y$ : $1 < b$
- if $x > y$, then $b^x < b^y$ : $0 < b < 1$

---

**MATHEMATICS LEVEL 2 Test - *Continued***

USE THIS SPACE FOR SCRATCH WORK.

13. What must be true about $x$ and $y$ if $\left(\frac{1}{2}\right)^{x-y} > 1$?

(A) $x > y$
(B) $x = y$
(C) $x < y$
(D) $xy > 0$
(E) $xy < 0$

14. How many real solutions are there for $2^x(x^2 - 4) = 0$?

(A) 0
(B) 1
(C) 2
(D) 3
(E) 4

## 3.5 Exponential Modeling

Compound interest rates, population growth(decay), or half-life modeling are all application of exponential modeling.

$$P_0(1+g)^t = P(t)$$

is the typical application of exponential model. However, there are some variations to this model.

- Compounded $n$ number of times in a year : $P_0(1+\frac{r}{n})^{nt}$ where $t$ is the number of years and $r$ is the annual interest rate. For instance, if it is compounded monthly, then the equation must be $P_0(1+\frac{r}{12})^{12t}$.
- Half-life equation is given by $P_0\left(\frac{1}{2}\right)^{\frac{t}{n}}$ where $t$ is the number of years and $n$ is the half-life. Half-life question may ask about the remaining amounts or the decayed amounts. Hence, read the question carefully.

---

**MATHEMATICS LEVEL 2 Test - *Continued***

15. If a Fair Bank has an annual interest rate of 3% compounded quarterly, and Bob invested \$1,000 as an initial deposit, how much would his savings account hold after 2 years?

USE THIS SPACE FOR SCRATCH WORK.

(A) 1,031
(B) 1,047
(C) 1,053
(D) 1,058
(E) 1,062

**MATHEMATICS LEVEL 2 Test - *Continued***

USE THIS SPACE FOR SCRATCH WORK.

16. If a group of scientists models that the half-life for bacteria known as "Slowly" is 10 years, and the current population is $2,510$, how many of the bacteria population will be gone after 15 years?

(A) 887.4
(B) 1261.5
(C) 1534.8
(D) 1622.6
(E) 2510

17. If the population of ant grows $0.11\%$ every year, and the current population is $1,520,000$, what is the percent increase of the population after 15 years?

(A) 0.017 percent
(B) 0.17 percent
(C) 1.7 percent
(D) 17 percent
(E) 170 percent

18. A new online bank takes two years to double the initial investment with the compound interest rate of

(A) 22.4 percent per year
(B) 26.8 percent per year
(C) 41.4 percent per year
(D) 58.6 percent per year
(E) 73.2 percent per year

## MATHEMATICS LEVEL 2 Test - *Continued*

USE THIS SPACE FOR SCRATCH WORK.

19. "Fraudulent" bank promised Bob to double his investment with the annual interest rate of 5% compounded monthly in certain number of years, but Bob forgot the exact number of years. Bob calculated it again later on and was deeply surprised at the number he found. What is the number of years, rounded up to the nearest integer, the bank offered Bob in the first place?

(A) 12
(B) 13
(C) 14
(D) 15
(E) 16

20. Downtrodden by the result, Bob moved to "Risky" bank to get the same promise, except that he forgot about the annual interest rate that is compounded monthly for five years to double the initial investment. He recalculated the interest rate at home, which is equal to

(A) 13.9 percent
(B) 24.1 percent
(C) 1.8 percent
(D) 75.9 percent
(E) 86.1 percent

## Answer Key to Practice Problems

1. (E)
2. (B)
3. (D)
4. (D)
5. (C)
6. (A)
7. (D)
8. (D)
9. (C)
10. (E)
11. (C)
12. (D)
13. (C)
14. (C)
15. (E)
16. (D)
17. (C)
18. (C)
19. (C)
20. (A)

# Detailed Solution for Practice Problems

1.

$$3^x = 27^{x-1}$$
$$3^x = (3^3)^{x-1}$$
$$3^x = 3^{3(x-1)}$$
$$3^x = 3^{3x-3}$$
$$x = 3x - 3$$
$$-2x = -3$$
$$x = \frac{-3}{-2}$$
$$x = 1.5$$

The reason why the exponents can be equalized is because the base is equal. Since the exponential function $y = 3^x$ is a $1-1$ function, especially strictly increasing function, the equal $y$-values imply equal $x$-values. This is the reason why that $a^m = a^m$ implies $m = n$ for $f(x) = a^x$. If there are two expressions with different bases without any common factor, then we must use *logarithm*. Similarly, putting logarithm in front of the two sides of the equation does not change the nature of one solution, since logarithmic function or exponential function are strictly increasing(or strictly decreasing). Hence, the answer is **(E)**.

2.

$$2^{2x} - 2^x - 2 = 0$$
$$(2^x)^2 - 2^x - 2 = 0$$
$$(2^x - 2)(2^x + 1) = 0$$
$$2^x = 2 \text{ or } -1$$
$$2^x = 2$$
$$x = 1$$

This is an application of quadratics in the form of exponential equation. Letting $2^x = X$, then we get $2^{2x} - 2^x - 2 = X^2 - X - 2 = 0$. Hence, $X = 2$ or $-1$, where $X = -1$ is an extraneous solution because any exponential expression cannot be negative. We call a solution *extraneous* if it is not counted as valid answer. Specifically, if $2^x = -1$, then $x$ cannot be real. Therefore, the answer is **(B)**.

**3.**

The first glance of the question tells us that the equation has different bases. However, if we look at the bases carefully, we notice that the power of 9 can be written as $9^x$, and it should be equal to $\frac{1}{27}$, so

$$\begin{aligned} 9^x &= \frac{1}{27} \\ 3^{2x} &= 3^{-3} \\ 2x &= -3 \\ x &= -\frac{3}{2} \end{aligned}$$

Hence, the answer must be **(D)**.

**4.**

$$\begin{aligned} f(t) &= 1{,}000{,}000 \\ 10^{2t-1} &= 10^6 \\ 2t-1 &= 6 \\ 2t &= 7 \\ t &= \frac{7}{2} \end{aligned}$$

The answer is **(D)**.

**5.**

Normally, when we solve a system of equations, we either subtract or add two equations together to eliminate one of the variables. On the other hand, if we have exponential expressions in the system of equations, we either multiply or divide two equations instead. In other words, this is an application of system of equations except that we either multiply or divide to eliminate the chosen variable.

$$\begin{cases} 2 = a \cdot 4^k \\ 4 = a \cdot 2^k \end{cases} \longrightarrow \frac{1}{2} = \frac{4^k}{2^k}$$

$$\longleftrightarrow \frac{1}{2} = 2^k$$

$$\longleftrightarrow 2^{-1} = 2^k$$

$$\longleftrightarrow -1 = k$$

Hence, the answer is **(C)**.

**6.**

Exponential expression in its primitive form is greater than 0. In other words, recall that $e^{-x} > 0$ for any real value $x$. Then,

$$\begin{aligned} e^{-x} &> 0 \\ 20e^{-x} &> 0 \\ 30 + 20e^{-x} &> 30 \\ y &> 30 \end{aligned}$$

Therefore, any value of $y$ less than or equal to 30 cannot be included in the range. Always remember the property that $a^x > 0$ for valid $a$. The answer must be **(A)**.

**7.**

What does the condition $a \neq b \implies f(a) \neq f(b)$ mean? It means that the function $y = f(x)$ is $1-1$ function. Different values of $x$ result in different outputs. How does it help us solve the question? Well, we are looking for functions that are either increasing or decreasing function.

- $y = \lfloor x \rfloor$ is not 1-to-1 function. For instance, $0 \leq x < 1$ are sent to 0.
- $y = x^2$ is 2-to-1 function. For instance, $x = -1$ and $x = 1$ are sent to $y = 1$.
- $y = |x|$ is 2-to-1 function. For instance, $x = 2$ and $x = -2$ are sent to $y = 2$.
- $y = e^{-x}$ is strictly decreasing.
- $y = x^{-2}$ is neither decreasing nor increasing. In fact, this is also 2-to-1 function. For instance, $x = 1$ and $x = -1$ are sent to $y = 1$.

The only function that matches with this condition is $y = e^{-x}$, i.e., **(D)**.

**8.**

Notice that $\frac{1}{a^{-1}} = a^{-(-1)} = a$. Hence,

$$\begin{aligned} a^2b^4c^3 &= \frac{9b^3c^3}{a^{-2}} \\ a^2b^4c^3 &= 9a^2b^3c^3 \\ b^4 &= 9b^3 \\ b &= 9 \end{aligned}$$

The answer is **(D)**.

**9.**

Since $x+y+z=9$, then $(x,y,z)=(2,3,4)$. However, this specific triple of $x$, $y$, and $z$ are not necessary. In fact,

$$\begin{aligned} a &= \sqrt[3]{2^x}\sqrt[3]{2^y}\sqrt[3]{2^z} \\ &= \sqrt[3]{2^x 2^y 2^z} \\ &= \sqrt[3]{2^{x+y+z}} \\ &= \sqrt[3]{2^9} \\ &= 2^3 \\ &= 8 \end{aligned}$$

Thus, the answer is **(C)**.

**10.**

$$\begin{aligned} 3^x + 3^x &= k3^{x+1} \\ 2\cdot 3^x &= k3^x 3 \\ 2\cdot 3^x &= (3k)3^x \\ 2 &= 3k \\ \frac{2}{3} &= k \end{aligned}$$

The answer is **(E)**.

**11.**

Notice that exponential function satisfies $f(a)f(b)=f(a+b)$ and $\frac{f(a)}{f(b)}=f(a-b)$. Suppose we do not know about the exponential function property. Then, let's apply the given condition to answer choices.

- $x-2=\frac{1}{4}x$ is NOT true for ALL $x$.
- $(x-2)^2=\frac{1}{4}x^2$ is NOT true for ALL $x$.
- $2^{x-2}=\frac{1}{4}2^x=\frac{2^x}{2^2}$ is true for ALL $x$.
- $\ln(x-2)=\frac{\ln(x)}{4}$ is NOT true for ALL $x$.
- $|x-2|=\frac{1}{4}|x|$ is NOT true for ALL $x$.

The only answer choice that satisfies the given condition is **(C)**.

**12.**

$$
\begin{aligned}
m^{2n} &= x \\
(m^{2n})^4 &= x^4 \\
m^{(2n)4} &= x^4 \\
m^{8n} &= x^4
\end{aligned}
$$

The answer is **(D)**.

**13.**

$$
\begin{aligned}
\left(\frac{1}{2}\right)^{x-y} &> 1 \\
\frac{1}{2^{x-y}} &> 1 \\
1 &> 2^{x-y} \\
2^0 &> 2^{x-y} \\
0 &> x-y \\
y &> x
\end{aligned}
$$

Be careful about the base value. If the base value is greater than 1, then $a^x > a^y$ is equivalent to $x > y$. However, if it is between 0 and 1, then $a^x > a^y$ is equivalent to $x < y$. Always make the base greater than 1, if possible. The answer must be **(C)**.

**14.**

According to zero-product property of real numbers,

$$\Box \times \triangle = 0 \text{ implies } \Box = 0 \text{ or } \triangle = 0.$$

Also, recall that $2^x > 0$ for all values of $x$. Use these two properties to solve for $x$.

$$
\begin{aligned}
2^x(x^2-4) &= 0 \\
x^2-4 &= 0 \\
x^2 &= 4 \\
x &= \pm 2
\end{aligned}
$$

The answer is **(C)**.

**15.**

This is a typical compound interest rate question. We must label $r$ as a decimal expression, not percentage. Since the annual interest rate is 3%, $r$ must be 0.03. However, the bank will compound the deposit four times in a year, so the quarterly interest rate must be $\frac{0.03}{4}$. Also, 2 years equal 8 quarters. Use these pieces of information, we will set up the equation. Let $f(t)$ be the amount Bob's savings account holds after $t$ years. Then,

$$f(t) = 1,000(1 + \frac{0.03}{4})^{4t}$$

Hence, $f(2) = 1,000(1 + \frac{0.03}{4})^{8} \approx 1,061.6 \approx 1,062$. The answer is **(E)**.

**16.**

This is a half-life modeling. The question asks us about the number of bacteria gone. The model, however, captures the size of the remaining population. In short, we must subtract the estimate from the original bacterial population. Let $S(t)$ be the amount of bacterial population after $t$ years such that

$$S(t) = 2,510 \times \left(\frac{1}{2}\right)^{\frac{t}{10}}$$

Since $S(15) = 2,510 \times \left(\frac{1}{2}\right)^{\frac{15}{10}} \approx 887.4$, so the bacterial population that will disappear equals $2,510 - 887.4 = 1,622.6$. The answer is **(D)**.

**17.**

First, be careful about the growth rate. 0.11% equals 0.0011 in decimal expression. Now, the population after 15 years equals $1,520,000 \times (1 + 0.0011)^{15} \approx 1,545,274$. In order to get the percent of increase from $A$ to $B$, recall that the equation is

$$\frac{\textbf{new} - \textbf{original}}{\textbf{original}} \times 100$$

Therefore, the percent increase of the population must be

$$\begin{aligned} \frac{1,545,274 - 1,520,000}{1,520,000} \times 100 &\approx 0.016627 \\ &\approx 1.6627\% \\ &\approx 1.7\% \end{aligned}$$

Hence, the answer is **(C)**.

**18.**

The bank must offer the annual interest rate of $r\%$, which is unknown. However, it takes only two years to double the original amount. Let's Let the initial amount in a new online bank be $P_0$, which stands for the principal amount or the initial amount. Then, the compound interest rate equation tells us that it takes 2 years with the annual interest rate of $r\%$, if

$$
\begin{aligned}
P_0\left(1+\frac{r}{100}\right)^2 &= 2P_0 \\
\left(1+\frac{r}{100}\right)^2 &= 2 \\
1+\frac{r}{100} &= \sqrt{2} \\
\frac{r}{100} &= \sqrt{2}-1 \\
r &= 100(\sqrt{2}-1) \\
r &\approx 41.42(\%)
\end{aligned}
$$

The answer is **(C)**.

**19.**

From the phrase *the annual interest rate of* 5% *compounded monthly*, we get $r = 0.05$, an annual interest rate, and the actual monthly rate is $\frac{0.05}{12}$. Let the initial amount Bob would like to invest is $B_0$ and $t$ be the number of years. Because $t$ is in years, the exponent must be $12t$, as it counts the number of months.

Also, as the exponent contains variable, we will use the change of base in logarithm, i.e.,

$$\log_a(b) = \frac{\log_\triangle(b)}{\log_\triangle(a)}$$

Hence,

$$
\begin{aligned}
B_0\left(1+\frac{0.05}{12}\right)^{12t} &= 2B_0 \\
\left(1+\frac{0.05}{12}\right)^{12t} &= 2 \\
12t &= \frac{\log(2)}{\log\left(1+\frac{0.05}{12}\right)} \\
t &= \frac{1}{12}\times\frac{\log(2)}{\log\left(1+\frac{0.05}{12}\right)} \\
t &\approx 13.89 \\
t &\approx 14
\end{aligned}
$$

Hence, the answer must be **(C)**.

**20.**

Let the initial amount Bob would like to invest is $B_0$ and the annual interest rate of $r$. Then,

$$\begin{aligned}
B_0\left(1+\frac{r}{1200}\right)^{12(5)} &= 2B_0 \\
\left(1+\frac{r}{1200}\right)^{60} &= 2 \\
1+\frac{r}{1200} &= \sqrt[60]{2} \\
\frac{r}{1200} &= \sqrt[60]{2}-1 \\
r &= 1200(\sqrt[60]{2}-1) \\
r &\approx 13.9(\%)
\end{aligned}$$

Therefore, the answer is **(A)**.

# CHECK ON LEARNING #2

## Definition of $e$

$$\lim_{n\to\infty}\left(1+\frac{1}{n}\right)^n = e$$

Imagine you went to a bank that guarantees to give you back $100\%$ after one year. Suppose you pu t a dollar in your savings account. The definition above shows that the limit value as $n$ grows infinitely large, then the number approaches $e$. This is a so-called *continuous compounding*, and the account value will approach $2.718\cdots\times 1$.

Let's have a look at the following example to compute $\lim_{n\to\infty}\left(1+\frac{1}{2n}\right)^n$. Always remember that

$$\lim_{\square\to\infty}\left(1+\frac{1}{\square}\right)^{\square} = e$$

Hence,

$$\begin{aligned}\lim_{n\to\infty}\left(1+\frac{1}{2n}\right)^n &= \lim_{2n\to\infty}\left(1+\frac{1}{2n}\right)^{(2n)(\frac{1}{2})}\\ &= \left(\lim_{n\to\infty}\left(1+\frac{1}{2n}\right)^{2n}\right)^{\frac{1}{2}}\\ &= e^{\frac{1}{2}}\\ &= \sqrt{e}\end{aligned}$$

In the equation above, $2n\to\infty$ implies that $n\to\infty$ with slower speed.

# Topic 4

# Logarithmic Functions and Expressions

- ✓ Arithmetic Properties of Logarithm
- ✓ Domain and Range Condition
- ✓ Inverse Function

## 4.1 Arithmetic Properties of Logarithm

The exam may directly ask you to find out whether the given properties are correct or use the properties to simplify the given expressions.

- $\log(xy) = \log(x) + \log(y)$
- $\log(x/y) = \log(x) - \log(y)$
- $\log(x^y) = y\log(x)$
- $\log(x)/\log(y) = \log_y(x)$
- $\log_x(x) = 1$
- $\log_x 1 = 0$
- $\log(0)$ does not exist.
- $\log(x) = \log_{10}(x)$
- $\ln(x) = \log_e(x)$

---

**MATHEMATICS LEVEL 2 Test - *Continued***

1. If $\log_a(2) = x$ and $\log_a(3) = y$, then $\log_a(24) =$ USE THIS SPACE FOR SCRATCH WORK.

(A) $3xy$
(B) $3+x+y$
(C) $3x+y$
(D) $x+3y$
(E) $3x-y$

2. If $\log_3(x) = -3$, then $x =$

(A) $-9$ (B) $\frac{1}{27}$ (C) $\frac{1}{9}$ (D) 27 (E) 9

## 4.2 Domain and Range Condition

First, let's have a look at domain.

- The base should be positive, not equal to 1.
- The input should be positive.

Because the input should be positive, the graph has a vertical asymptote at $x = 0$. The parent function $y = \log_b(x)$ has no $y$-intercept, but the graph of the transformed function may have a $y$-intercept.

On the other hand, the range is all real numbers. This is not only 1-to-1 function, but also "onto"[1] function.

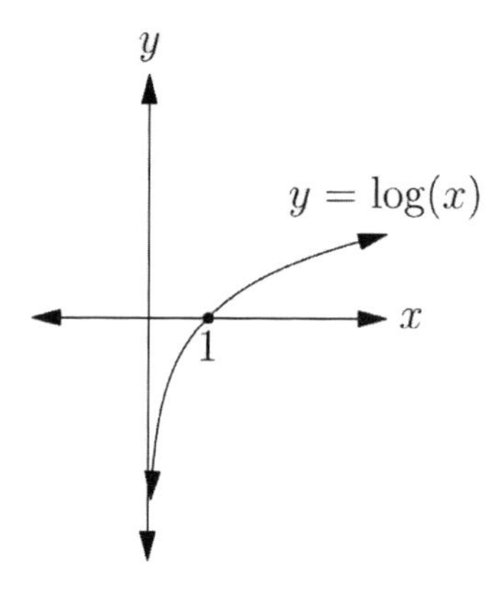


Figure of $y = \log(x)$

---

**MATHEMATICS LEVEL 2 Test - *Continued***

3. Which of the following function has a vertical asymptote at $x = 1$?

(A) $y = \dfrac{x^2 - 1}{x - 1}$

(B) $y = \dfrac{1}{x^2 + 1}$

(C) $y = e^{x-1}$

(D) $y = \ln(x - 1)$

(E) $y = (x - 1)^2$

USE THIS SPACE FOR SCRATCH WORK.

[1] We call such function as surjective function, and usually the function is *onto* the set of real numbers if the range is the set of all real numbers.

**MATHEMATICS LEVEL 2 Test - *Continued***

USE THIS SPACE FOR SCRATCH WORK.

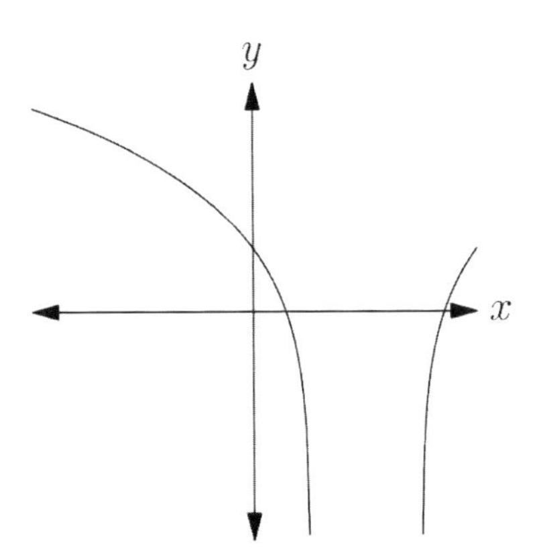


Figure of $y = \log(x^2 - 4x + 3)$

4. Which of the following values of $x$ is undefined for the function $f(x) = \log(x^2 - 4x + 3)$?

(A) 0
(B) 2
(C) 4
(D) 6
(E) 8

5. Which of the following values of $x$ is not included in the domain of $y = \log_{(7-x)}(x - 1)$?

(A) 1
(B) 2
(C) 3
(D) 4
(E) 5

In order to check whether a given function has an inverse function or not, we draw any horizontal line to see if it intersects the graph with more than one point. If so, the function does not have an inverse; otherwise, it passes the test so that it has an inverse function.

Logarithmic function and exponential function are typical examples that pass the inverse function. Even more, they are the examples of bijective function, meaning that they are one-to-one and surjective.

---

**MATHEMATICS LEVEL 2 Test - *Continued***

6. Which of the following functions does not have an inverse function?

USE THIS SPACE FOR SCRATCH WORK.

(A) $y = \sin(x)$

(B) $y = \ln(x)$

(C) $y = e^x$

(D) $y = -x$

(E) $y = \frac{1}{x}$

7. A function $f$ is known to be bijective if $f$ is 1-to-1 and onto. Which of the following functions is bijective?

(A) $y = \cos(x)$

(B) $y = \sin(x)$

(C) $y = \frac{1}{x^2}$

(D) $y = \ln(x)$

(E) $y = \tan(x)$

## 4.3 Inverse Function

First off, inverse function undoes what the original function does.

$$f(a) = b \qquad\qquad f^{-1}(b) = a$$

In order to have an inverse function, the function must be 1-to-1, passing the Horizontal Line Test, abbreviated as HLT, which was referenced in the previous section. The procedure of the test begins by drawing multiple horizontal lines and see if there is at most one point of intersection. Otherwise, it fails, and there is no inverse function. The following properties of inverse function are useful when we solve the problems in SAT Math Level 2.

- $f(f^{-1}(x)) = x$ and $f^{-1}(f(x)) = x$, where $x$ is in the domain of the inner function.
- $f(x) = f^{-1}(x)$ satisfies $f(f(x)) = x$.

Specifically, when we compose logarithmic function and exponential function, two of which are inverse to each other, then

- $e^{\ln(x)} = x$ and $\ln(e^x) = x$.
- $a^{\log_a(x)} = x$ and $\log_a(a^x) = x$.

Let's have a look at $f(x) = f^{-1}(x)$. Since the function has **self-inverse**, meaning that its inverse equals itself, the graph of $y = f(x)$ is symmetric about $y = x$.

---

**MATHEMATICS LEVEL 2 Test** - ***Continued***

8. If $f(x) = 3 - \log_2(x)$ for $x > 0$, then $f^{-1}(4) =$

USE THIS SPACE FOR SCRATCH WORK.

(A) $\sqrt{2}$

(B) $\frac{1}{2}$

(C) $\frac{1}{\sqrt{2}}$

(D) 2

(E) 4

**MATHEMATICS LEVEL 2 Test - *Continued***

USE THIS SPACE FOR SCRATCH WORK.

9. If $f(x) = f^{-1}(x)$, which of the following must be true?

(A) Its graph is symmetric about the $x$-axis.
(B) Its graph is symmetric about the $y$-axis.
(C) Its graph is symmetric about the origin.
(D) Its graph is symmetric about $y = x$.
(E) Its graph is symmetric about $y = -x$.

10. For any real value $x$, $2e^{\ln(2x+4)} =$

(A) $4x+8$
(B) $2x+4$
(C) $4x+2$
(D) $2x+8$
(E) $8x+4$

11. If $f(x) = \ln(ex)$ and $g(x) = e^{x-7}$, then $(f \circ g)(2) =$

(A) $-5$
(B) $-4$
(C) 2
(D) 4
(E) 5

## MATHEMATICS LEVEL 2 Test - *Continued*

USE THIS SPACE FOR SCRATCH WORK.

12. Which of the following satisfies $f(x) = f^{-1}(x)$?

(A) $y = \ln(x)$
(B) $y = x^2$
(C) $y = |x|$
(D) $y = -x$
(E) $y = e^x$

13. Which of the following functions has the inverse function with the largest slope?

(A) $y = x$
(B) $y = 2x$
(C) $y = 3x$
(D) $y = 4x$
(E) $y = 5x$

14. All of the following functions have an inverse function EXCEPT

(A) $y = \ln(x)$
(B) $y = \tan^{-1}(x)$
(C) $y = |x|$
(D) $y = \sqrt[3]{x}$
(E) $y = e^x$

**MATHEMATICS LEVEL 2 Test - *Continued***

15. Which of the following functions satisfies $f(f(x)) = x$?

USE THIS SPACE FOR SCRATCH WORK.

I. $f(x) = 4 - x$

II. $f(x) = -\frac{1}{x}$

III. $f(x) = 2x - 4$

(A) I only
(B) II only
(C) I and II
(D) II and III
(E) I, II, and III

## Answer Key to Practice Problems

1. (C)

2. (B)

3. (D)

4. (B)

5. (A)

6. (A)

7. (D)

8. (B)

9. (D)

10. (A)

11. (B)

12. (D)

13. (A)

14. (C)

15. (C)

## Detailed Solution for Practice Problems

**1.**

Recall that $\log(\square \times \triangle) = \log(\square) + \log(\triangle)$. Also, $\log(\square^{\triangle}) = \triangle \times \log(\square)$. Therefore,

$$\begin{aligned}\log_a(24) &= \log_a(2^3 \cdot 3^1)\\ &= \log_a(2^3) + \log_a(3^1)\\ &= 3\log_a(2) + 1\log_a(3)\\ &= 3x + y\end{aligned}$$

The answer must be **(C)**.

**2.**

Logarithmic form and exponential form can be interchanged as $\square^{\triangle} = \bigstar \implies \log_{\square}\bigstar = \triangle$.

$$\begin{aligned}\log_3(x) &= -3\\ 3^{-3} &= x\\ \frac{1}{3^3} &= x\\ \frac{1}{27} &= x\end{aligned}$$

Hence, the answer is **(B)**.

**3.**

- (A) has no vertical asymptote, but hole at $x = 1$.
- (B) has no vertical asymptote.
- (C) has a horizontal asymptote at $y = 0$.
- (D) has a vertical asymptote at $x = 1$.
- (E) has no vertical asymptote.

Therefore, the answer is **(D)**.

**4.**

$f(x) = \log(x^2 - 4x + 3)$ must satisfy $x^2 - 4x + 3 > 0$. In other words, $x^2 - 4x + 3 = (x-3)(x-1) > 0$, so $x < 1$ or $3 < x$. At $x = 2$, the function is undefined, and as one can see from the figure of $y = \log(x^2 - 4x + 3)$, the function is undefined at $x = 2$, so the answer is **(B)**.

**5.**

Given $y = \log_{7-x}(x-1)$, the base must satisfy

- $7 - x > 0$
- $7 - x \neq 1$

Similarly, the input must satisfy $x - 1 > 0$. Hence, $1 < x < 7$ where $x \neq 6$. The only value that does not fit in the inequality from the answer choices is **(A)**.

**6.**

The inverse function exists if and only if the original function is $1-1$ function. Other than (A), all the other functions from (B) to (E) are $1-1$. Hence, the answer is **(A)**.

**7.**

Out of all answer choices, the only function that is both injective and surjective is $y = \ln(x)$. In fact, all the other functions are not $1-1$, i.e., injective. Hence, the answer is **(D)**.

**8.**

Let $f^{-1}(4) = x$. Then,

$$\begin{aligned} 4 &= 3 - \log_2(x) \\ -1 &= \log_2(x) \\ 2^{-1} &= x \end{aligned}$$

Hence, the answer is **(B)**.

**9.**

This function has self-inverse, which means that the graph of $y = f(x)$ is symmetric about the line $y = x$. Hence, the answer must be **(D)**.

**10.**

Recall that $e^{\ln(\bigstar)} = \bigstar$. Therefore,

$$\begin{aligned} 2e^{\ln(2x+4)} &= 2(2x+4) \\ &= 4x+8 \end{aligned}$$

The answer is **(A)**.

**11.**

$$\begin{aligned}(f \circ g)(2) &= f(g(2)) \\ &= f(e^{2-7}) \\ &= f(e^{-5}) \\ &= \ln(e \cdot e^{-5}) \\ &= \ln(e^{-4}) \\ &= -4\ln(e) \\ &= -4\end{aligned}$$

The answer is **(B)**.

**12.**

If $f(x) = f^{-1}(x)$, then $f(f(x)) = x$. First, (B) and (C) should be eliminated because they do not have inverse functions. (A) and (E) are eliminated because $y = \ln(x)$ and $y = e^x$ are inverse to each other. Therefore, the answer must be **(D)**.

**13.**

The inverse function of $y = kx$ is $y = \frac{1}{k}x$. Therefore, the inverse function has the largest slope if $\frac{1}{k}$ is largest. Out of all answer choices, the inverse function for $y = x$ has the largest slope. Hence, the answer is **(A)**.

**14.**

The function $y = |x|$ has no inverse function because it is 2-to-1 function, meaning that 2 inputs are connected to 1 output. Hence, the answer is **(C)**.

**15.**

I. $f(f(x)) = 4 - (4 - x) = x$.

II. $f(f(x)) = -\dfrac{1}{-1/x} = x$.

III. $f(f(x)) = 2(2x - 4) - 4 = 4x - 8 - 4 = 4x - 12 \neq x$.

Therefore, I and II are true. The answer must be **(C)**.

# CHECK ON LEARNING #3

## Inverse Relation between $e^x$ and $\ln(x)$

$$e^{\ln(x)} = x \text{ and } \ln(e^x) = x$$

Let $f(x) = e^x$ and $f^{-1}(x) = \ln(x)$. Then, by the composition of inverse functions,

$$\begin{aligned} f(f^{-1}(x)) &= f(\ln(x)) \\ &= e^{\ln(x)} \\ &= x \end{aligned}$$

On the other hand, let $f^{-1}(f(x)) = x$. By the composition of inverse functions,

$$\begin{aligned} f^{-1}(f(x)) &= \ln(f(x)) \\ &= \ln(e^x) \\ &= x \end{aligned}$$

This inverse function property shows that

$$\square^{\log_{\square}(\triangle)} = \triangle \qquad \log_{\square}(\square^{\triangle}) = \triangle$$

1. Application of $\square^{\log_{\square}(\triangle)} = \triangle$

$$\begin{aligned} 9^{\log_3(5)} &= (3^2)^{\log_3(5)} \\ &= 3^{2\log_3(5)} \\ &= 3^{\log_3(25)} = 25 \end{aligned}$$

2. Application of $\log_{\square}(\square^{\triangle}) = \triangle$

$$\begin{aligned} \log_2 8^3 &= \log_2 (2^3)^3 \\ &= \log_2(2^9) \\ &= 9 \end{aligned}$$

# Topic 5

# Polynomial Functions and Expressions

✓ Graph Analysis

✓ Factor Theorem and Remainder Theorem

✓ Range of Polynomial Function

✓ Zeros of Polynomials

✓ Intermediate Value Theorem

## 5.1 Graph Analysis

Normally, either using a graphic calculator or not, we use the intercept analysis. Let's have a look at five major themes of graph analysis.

- Multiplicity is either even or odd. If $(x-a)$ has **even** multiplicity, then the graph is **tangent** at $x=a$. **Otherwise**, the graph **cuts through** $x=a$.
- Degree is either even or odd. If the degree is **even**, the end-behavior is either **up-up** or **down-down**. **Otherwise**, the end-behavior is either **up-down** or **down-up**.
- $y$-**intercept** is the **constant term**.
- Maximum and minimum over the closed interval of $x$ is usually found by graphing the function.
- Increasing or decreasing function can easily determined by graphing the function.

---

**MATHEMATICS LEVEL 2 Test - *Continued***

USE THIS SPACE FOR SCRATCH WORK.

1. What is the minimum number of $x$-intercepts for $y=ax^7+bx^3+c$ for $a\neq 0$?

(A) 0
(B) 1
(C) 2
(D) 3
(E) 4

2. Given $y=(x-a)(x-b)(x-c)(x-d)$, and the graph has two $x$-intercepts, which of the following could be true?

(A) $a=b=c\neq d$
(B) $a=b\neq c=d$
(C) $a=b\neq c\neq d$
(D) $a\neq b\neq c\neq d$
(E) None of the above

## 5.2 Factor Theorem and Remainder Theorem

Remainder theorem(or factor theorem) appears quite frequently in the exam. Let's have a look at what they are.

$$\underbrace{P(x)}_{\text{Dividend}} = \underbrace{(x-a)}_{\text{Divisor}} \times \underbrace{Q(x)}_{\text{Quotient}} + \underbrace{P(a)}_{\text{Remainder}}$$

where $P(a) = r$, known as a remainder. This long-division algorithm of polynomial division is used for remainder theorem or factor theorem.

- **The remainder when $P(x)$ is divided by $x - a$** equals $P(a)$.
- If $P(a) = 0$, then $x - a$ **must be a factor of** $P(x)$ because the remainder is 0.

---

**MATHEMATICS LEVEL 2 Test - *Continued***

USE THIS SPACE FOR SCRATCH WORK.

3. If $x - 1$ is a factor of $x^3 + 2x^2 + kx + 3$, which of the following must be the value of $k$?

(A) −5
(B) −6
(C) 1
(D) 6
(E) 5

4. If $f(x) = (2 - x)g(x) + r$, then $r =$

(A) $g(2)$
(B) 2
(C) $f(2)$
(D) 0
(E) $f(2)g(2)$

## 5.3 Range of Polynomial Functions

The range of $f(x) = a_n x^n + a_{n-1}x^{n-1} + \cdots + a_0$ depends on the parity (even/oddness) of $n$.

- $n$ is odd : the range is $\mathbb{R}$, the set of all real numbers.
- $n$ is even : the range must be carefully computed by the graphic calculator.

However, everything changes if the domain is restricted. If that happens, graph the function on the graphic calculator to plot either maximum or minimum value.

---

**MATHEMATICS LEVEL 2 Test - *Continued***

USE THIS SPACE FOR SCRATCH WORK.

5. If $-1 \leq x \leq 2$, the maximum value of $f(x) = -2x^5 + 3x - 7$ is

(A) $-8.78$
(B) $-5.22$
(C) $-4.87$
(D) $-0.75$
(E) 0.74

6. What is the minimum number of $x$-intercepts for $y = ax^8 + bx^6 + cx^4 + d$ for $a \neq 0$?

(A) 0
(B) 1
(C) 2
(D) 3
(E) 4

## 5.4 Zeros of Polynomial Functions

Most polynomials (I think, all polynomials) in SAT Math Level 2 have real coefficients, sometimes integer coefficients. In this case, by the Fundamental Theorem of Algebra, if a conjugate form $a+bi$ or $a+\sqrt{b}$ is given as a zero of polynomial, then the other conjugate form is automatically the zero as well.

- If $a \pm bi$ is a root of $p(x)=0$ for polynomial $p(x)$ with real coefficients, then $a \mp bi$ is also a root.
- If $a \pm \sqrt{b}$ is a root of $p(x)=0$ for polynomial $p(x)$ with rational coefficients, then $a \mp \sqrt{b}$ is also a root.

There is another way to solve this type of questions. We simply let $x=a+bi$ and rearrange the terms. We will have a look at the following practice questions.

Also, there is a generalized version of Vieta's formula. Typically, this formula is extended upto cubic polynomial equations. Let $r, s, t$ be the zeros of $ax^3+bx^2+cx+d=0$, then

- $-\frac{b}{a}=r+s+t$
- $\frac{c}{a}=rs+st+tr$
- $-\frac{d}{a}=rst$

---

**MATHEMATICS LEVEL 2 Test - *Continued***

7. If $-2+\sqrt{3}$ is a zero of a polynomial function with integer coefficients, which of the following must also be a zero?

(A) $2-\sqrt{3}$
(B) $2+\sqrt{3}$
(C) $3-\sqrt{2}$
(D) $3+\sqrt{2}$
(E) $-2-\sqrt{3}$

USE THIS SPACE FOR SCRATCH WORK.

**MATHEMATICS LEVEL 2 Test - *Continued***

USE THIS SPACE FOR SCRATCH WORK.

8. If $1+i$ is zero of the polynomial function $f(x)$, then which of the following must be a factor of $f(x)=0$?

(A) $x^2-2x+1$
(B) $x^2+x+1$
(C) $x^2-x+1$
(D) $x^2-2x+2$
(E) $x^2+2x+2$

9. If $x^3+x+1=0$, what must be true about the zeros?

(A) The product of all zeros is 1.
(B) The sum of all zeros is 1.
(C) The product of all zeros is 0.
(D) The sum of all zeros is 0.
(E) The sum of all zeros is $-1$.

10. If the difference of the integer roots of $x^2+ax+7=0$ is 8, then which of the following is the sum of all possible values of $a$?

(A) $-6$
(B) 6
(C) 0
(D) 7
(E) $-7$

## 5.5 Intermediate Value Theorem

This theorem, although it sounds difficult and challenging, will come out in the test because all polynomial functions are continuous.

In SAT Math Level 2, it is safe to say that all continuous functions have connected graphs. If graphs are connected, then a bizarre phenomenon happens in the graph.

Assume $a < b$ and $f(a) < 0$ and $f(b) > 0$, for a continuous function $f(x)$. Then, there is at least one real zero between $x = a$ and $x = b$. Think about drawing the graph without making any gap in the graph. Then, the graph must pass through the $x$-axis at a certain point between $x = a$ and $x = b$! This is the intermediate value theorem in application.

The function $f(x)$ is continuous at $[a, b]$ where $f(a)f(b) < 0$, meaning that $f(a) < 0$, $f(b) > 0$ or $f(a) > 0$, $f(b) < 0$. Then, there must be at least one real zero between $x = a$ and $x = b$.

---

**MATHEMATICS LEVEL 2 Test - *Continued***

11. A polynomial function $f(x)$ passes through $(1, -2)$, $(2, 3)$ and $(3, 6)$ Which of the following must be true?

(A) There is at least one zero between $x = 1$ and $x = 2$.
(B) There is at least one zero between $x = 2$ and $x = 3$.
(C) The function $f(x)$ is increasing for $1 < x < 2$.
(D) The function $f(x)$ is increasing for $2 < x < 3$.
(E) The function $f(x)$ is increasing for $1 < x < 3$.

USE THIS SPACE FOR SCRATCH WORK.

## MATHEMATICS LEVEL 2 Test - *Continued*

USE THIS SPACE FOR SCRATCH WORK.

| $x$ | $f(x)$ |
|---|---|
| $-1$ | 2 |
| 0 | 0 |
| 1 | $-2$ |
| 2 | 1 |

12. If $f(x)$ is a continuous function, which of the following must be true?

(A) The function $f(x)$ is decreasing for $-1 < x < 0$.
(B) The function $f(x)$ is increasing for $1 < x < 2$.
(C) There is at least one zero between $x = 1$ and $x = 2$.
(D) There is at least one zero between $x = -1$ and $x = 0$.
(E) There is exactly one zero between $x = -1$ and $x = 1$.

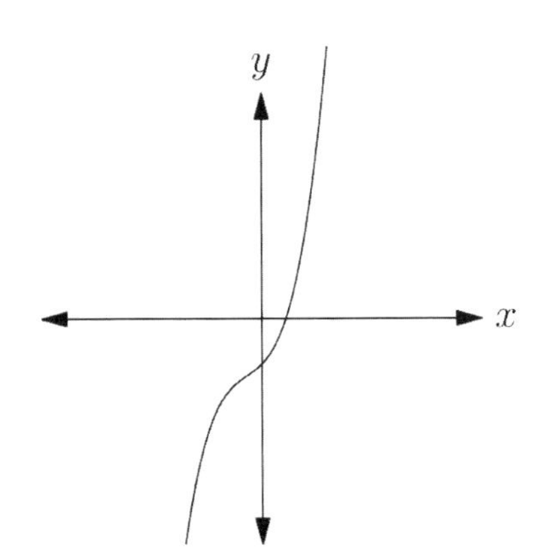


13. On which interval does the function $f(x) = x^3 + x^2 + x - 1$ have at least one $x$-intercept?

(A) $(-1, 0)$
(B) $(0, 1)$
(C) $(1, 2)$
(D) $(2, 3)$
(E) $(3, 4)$

**MATHEMATICS LEVEL 2 Test - *Continued***

USE THIS SPACE FOR SCRATCH WORK.

14. Which of the following has no $x$-intercept?

(A) $y = x^3 - 1$
(B) $y = \sin(x)$
(C) $y = \cos(x)$
(D) $y = \tan(x)$
(E) $y = \sec(x)$

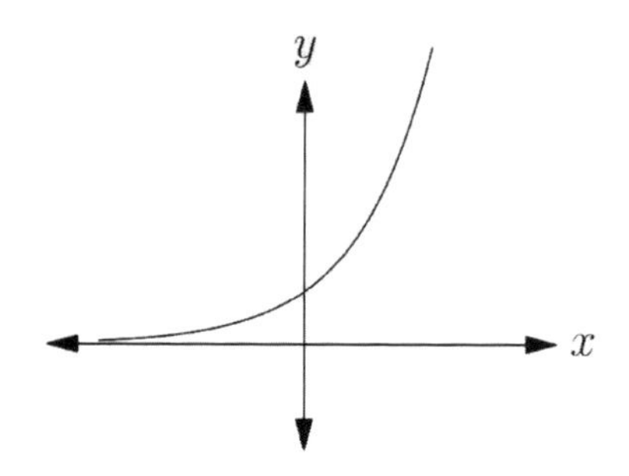


15. Which of the following must be true about $y = e^x$?

(A) $e^x = 2$ for some $x$ between 0 and 1.
(B) $y = e^x$ is decreasing for some positive $x$.
(C) There is at least one $x$-intercept between $-1$ and 0.
(D) The graph is concave down.
(E) There is a vertical asymptote at $x = 0$.

## Answer Key to Practice Problems

1. (B)

2. (B)

3. (B)

4. (C)

5. (B)

6. (A)

7. (E)

8. (D)

9. (D)

10. (C)

11. (A)

12. (C)

13. (B)

14. (E)

15. (A)

## Detailed Solution for Practice Problems

**1.**

Look at the degree 7. If it is odd, then the end-behavior must be either up-down or down-up. This should result in the existence of at least one $x$-intercept. Therefore, the answer is **(B)**.

**2.**

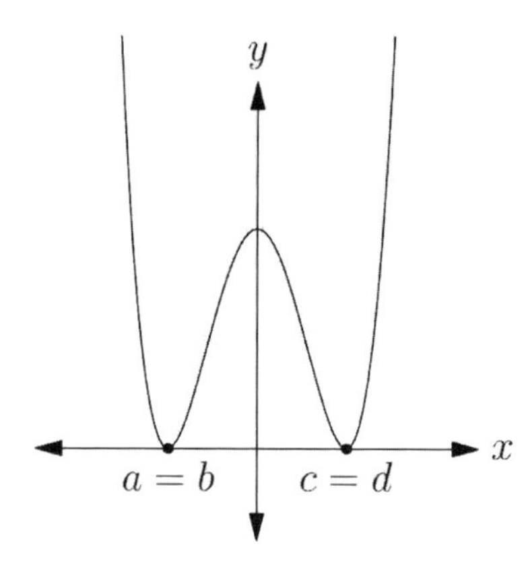


If the polynomial function of degree four has two distinct $x$-intercepts, then the graph must look like the figure above. As one can see, out of all four $a, b, c, d$'s, there are only two distinct numerical values, meaning that two letters of which should be equal to one another. Hence, the answer is **(B)**.

**3.**

By Factor Theorem, $x-1$ is a factor of $x^3+2x^2+kx+3$ if and only if $1^3+2(1)^2+k(1)+3=0$. Hence, $1+2+k+3=0$, so $k=-6$. The answer is **(B)**.

**4.**

This is an application of remainder theorem. Simply substitute the zero of the divisor into the whole expression and see what happens.

$$\begin{aligned} f(2) &= (2-2)g(2)+r \\ &= 0g(2)+r \\ &= r \end{aligned}$$

The answer must be **(C)**.

5.

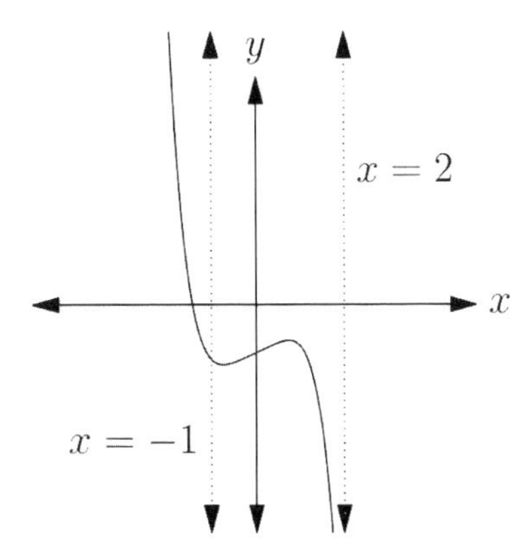


As one can see from the figure above, the maximum value of $y = -5.22$ occurs at $x = 0.74$. Hence, the answer is **(B)**.

6.

Similar to question 1, look at the degree 8. If the degree is even, then the existence of $x$-intercept is not guaranteed. Hence, the answer is **(A)**.

7.

If a polynomial function has integer coefficients and $a+\sqrt{b}$ is its root, then $a-\sqrt{b}$ is also one of the roots. Therefore, if $-2+\sqrt{3}$ is a zero of a certain polynomial function with integer coefficients, then $-2-\sqrt{3}$ is also a zero. Therefore, the answer is **(E)**.

8.

If $1+i$ is a zero, then $1-i$ is also zero. Therefore, $(x-(1+i))(x-(1-i))$ should be a factor of $f(x)$. So,

$$\begin{aligned}(x-(1+i))(x-(1-i)) &= x^2-(1+i+1-i)x+(1-i)(1+i)\\ &= x^2-2x+(1^2-i^2)\\ &= x^2-2x+2\end{aligned}$$

The answer is **(D)**.

**9.**

The Vieta's formula tells us about $x^3+0x^2+x+1=(x-r)(x-s)(x-t)=0$ such that

- $r+s+t=-\frac{0}{1}=0$
- $rs+st+rt=\frac{1}{1}=1$
- $rst=-\frac{1}{1}=-1$

The answer must be **(D)**.

**10.**

Let's denote the roots by $r$ and $s$, for the sake of convenience. The condition that the product of $r$ and $s$ is 7 implies that $(r,s)=(1,7),(-1,-7),(7,1),(-7,-1)$. Hence, the values of $a$ are 8 or $-8$. The sum of values of $a$ is $8+(-8)=0$, so the answer is **(C)**.

**11.**

A polynomial function is continuous so its graph is connected. It means that there should be at least one $x$-intercept between $x=1$ and $x=2$ because $f(1)=-2<0$ and $f(2)=3>0$, by so-called intermediate value theorem. The answer is **(A)**.

**12.**

If you see a phrase *continuous*, it is likely that the question asks you about the IVT. Circle the part where the sign of $y$-value changes. At $x=1$, $y=-2<0$. However, at $x=2$, $y=1>0$. Therefore, according to the intermediate value theorem, there should be at least one $x$-intercept between $x=1$ and $x=2$. The answer should be **(C)**.

**13.**

At $x=0$, $f(0)=0^3+0^2+0-1=-1<0$. However, at $x=1$, $f(1)=1^3+1^2+1-1=2>0$. By IVT, $y=f(x)$ has at least one $x$-intercept between $x=0$ and $x=1$. Therefore, the answer is **(B)**.

**14.**

(A) is false because it is polynomial with degree 3, meaning that there is at least one $x$-intercept. (B) and (C) are false because they are continuous function with the range between $[-1,1]$, so IVT states that there are $x$-intercepts. (D) is false because $y=\tan(x)$ has at least one $x$-intercept, i.e., $x=0$. On the other hand, $y=\sec(x)$ is discontinuous function whose range is $(-\infty,-1]\cup[1,\infty)$, implying that there is no $x$-intercept. The answer is **(E)**. The figure below shows the graph of $y=\sec(x)$, and indeed there is no $x$-intercept.

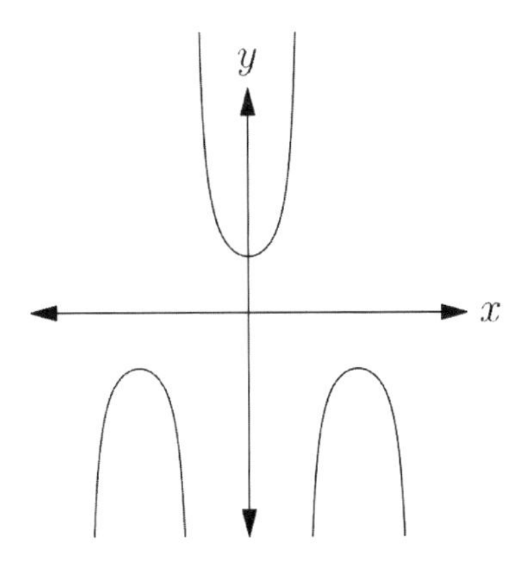


**15.**

At $x=0$, $y=e^0=1$. However, at $x=1$, $y=e^1=e>2$. Therefore, there is at least one $x$-value between $x=0$ and $x=1$ such that $e^x=2$, by IVT. The answer is **(A)**. Have a look at the following graph and try to make sense out of it.

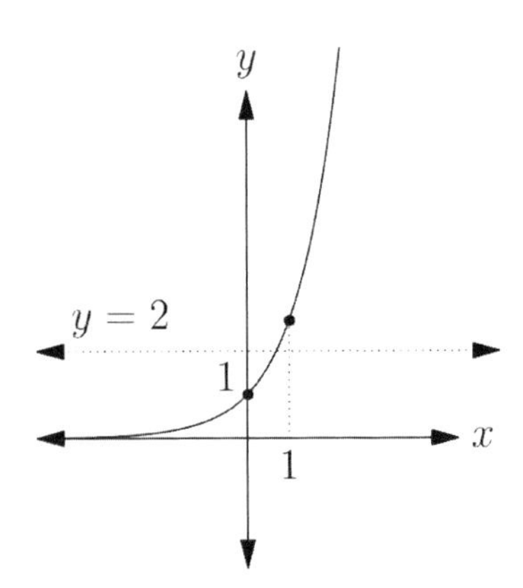

# Topic 6

# Piecewise Functions and Expressions

- ✓ Piecewise Expressions
- ✓ Floor Function
- ✓ Absolute-value Function
- ✓ Finding Equivalent Expressions

## 6.1 Piecewise Expressions

The key idea for piecewise expressions is to correctly substitute values in valid domain. There are generally two or three subdomains, for instance,

$$f(x) = \begin{cases} p(x) & \text{if } x \leq a \\ q(x) & \text{if } a < x \leq b \\ \underbrace{r(x)}_{\text{Functions}} & \text{if } \underbrace{b \leq x}_{\text{Subdomains}} \end{cases}$$

**MATHEMATICS LEVEL 2 Test - *Continued***

USE THIS SPACE FOR SCRATCH WORK.

1. If a function $f(x)$ is defined by

$$f(x) = \begin{cases} 2 & x < 1 \\ 3 & 1 \leq x < 3 \\ 4 & 3 \leq x \end{cases}$$

which of the following is $f(2)$?

(A) 1 (B) 2 (C) 3 (D) 4 (E) 5

2. If a function $f(x)$ is defined by

$$f(x) = \begin{cases} |x| & -1 < x < 1 \\ 1 & \text{otherwise} \end{cases}$$

which of the following cannot be in its range?

(A) 0 (B) $\frac{1}{3}$ (C) $\frac{1}{2}$ (D) 1 (E) 2

## MATHEMATICS LEVEL 2 Test - *Continued*

USE THIS SPACE FOR SCRATCH WORK.

3. If a function $f(x)$ is defined by

$$f(x) = \begin{cases} 1 & x \text{ is even} \\ -1 & x \text{ is odd} \\ 0 & x \text{ is neither} \end{cases}$$

which of the following is the period[1] of $f(x)$?

(A) 0
(B) 1
(C) 2
(D) 3
(E) 4

4. If a function $f(x)$ is defined by

$$f(x) = \begin{cases} 2x+b & x < 3 \\ 3x-1 & 3 \leq x \end{cases}$$

what value of $b$ makes the function continuous?

(A) $-2$
(B) $-1$
(C) 0
(D) 1
(E) 2

[1] The period of function is the value of $k$ when $f(x+k) = f(x)$ for all $x$ in its domain. Think about the smallest horizontal length of the graph that you should replicate to make the whole graph.

## 6.2 Floor Function

Floor function is a typical example of piecewise function. It is denoted by

$$f(x) = \lfloor x \rfloor = [x]$$

where $\lfloor x \rfloor$ is the greatest integer smaller than or equal to $x$.[2]

- if $n \leq x < n+1$, then $\lfloor x \rfloor = n$.
- The domain of $f(x) = \lfloor x \rfloor$ is the set of all real numbers.
- The range of $f(x) = \lfloor x \rfloor$ is the set of all integers.
- Its graph is given by the steps, i.e.,

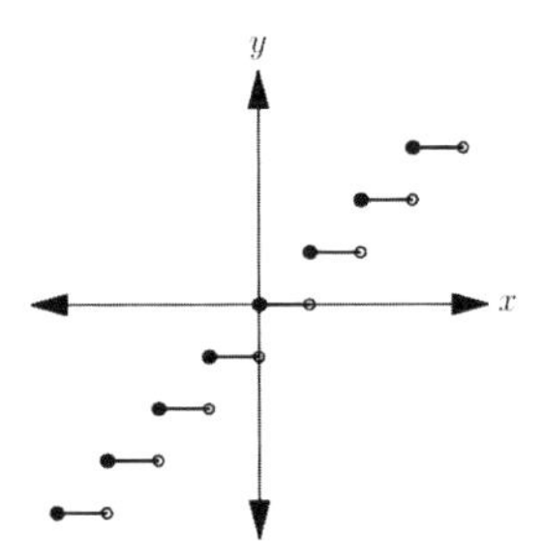


**MATHEMATICS LEVEL 2 Test -** ***Continued***

5. If $f(x) = \lfloor \lfloor x \rfloor \rfloor$, what is the value of $f(-1.46)$? USE THIS SPACE FOR SCRATCH WORK.

(A) $-1$
(B) $-2$
(C) 0
(D) 2
(E) 1

[2]There is a corresponding concept, namely the ceiling of $x$, denoted by $\lceil x \rceil$, which is the least integer greater than or equal to $x$. Floor functions will appear more often the ceiling functions, and the question will specify the condition of $x$ right next to the notation, so don't worry about notational confusions.

**MATHEMATICS LEVEL 2 Test - *Continued***

USE THIS SPACE FOR SCRATCH WORK.

6. $\lfloor -2.45 \rfloor + \lfloor 2.44 \rfloor =$

(A) 0
(B) 1
(C) $-1$
(D) $-0.01$
(E) None of the above

7. What is the line of symmetry for $y = |\lfloor x \rfloor|$?

(A) $x = 0$
(B) $x = \frac{1}{2}$
(C) $x = 1$
(D) $x = -\frac{1}{2}$
(E) $x = 2$

8. What is the range of $y = x - \lfloor x \rfloor$?

(A) $[0, 1]$
(B) $(0, 1)$
(C) $[0, 1)$
(D) $(0, 1]$
(E) 0 and 1

## 6.3 Absolute-value Function

$$f(x) = |x|$$

$|\cdot|$ is the symbol for absolute value. Absolute-value function is another application of piecewise function with two subdomains.

$$f(x) = |x| = \begin{cases} x & 0 \le x \\ -x & x < 0 \end{cases}$$

This is a line graph whose vertex is at $(0,0)$, meaning that

- $x = 0$ is the axis of symmetry.
- $y = 0$ is the minimum value of $y = f(x)$.

In SAT Math Level 2, it is so important to get rid of the absolute-value expression using the following rules.

$$|a-b| = \begin{cases} a-b & 0 \le a-b \\ b-a & a-b < 0 \end{cases}$$

---

**MATHEMATICS LEVEL 2 Test - *Continued***

9. If $f(x-1) = |4x-3|$, then $f(3) =$

USE THIS SPACE FOR SCRATCH WORK.

(A) 13 (B) 14 (C) 15 (D) 16 (E) 17

10. For $a < b$, $|a-b| - |b-a| =$

(A) $a-b$
(B) $2(a-b)$
(C) 0
(D) $2(b-a)$
(E) $b-a$

**MATHEMATICS LEVEL 2 Test - *Continued***

USE THIS SPACE FOR SCRATCH WORK.

11. $\lim\limits_{x \to 2} \dfrac{|x-2|}{x-2} =$

(A) 2
(B) 1
(C) $-1$
(D) $-2$
(E) Undefined

12. What value of $x$ does the graph of $y = -|4-x|+3$ reach its maximum?

(A) 3
(B) 4
(C) 0
(D) $-4$
(E) $-3$

13. Which of the following satisfies $|x|+|y| < 4$?

(A) $(4,0)$
(B) $(-3,1)$
(C) $(2,4)$
(D) $(-1,-2)$
(E) $(-4,0)$

## 6.4 Finding Equivalent Expressions

The following expression is one of the most common expressions you would find during the test.

$$\sqrt{x^2} = |x|$$

**MATHEMATICS LEVEL 2 Test -** ***Continued***

USE THIS SPACE FOR SCRATCH WORK.

14. If $r > 0$, which of the following is equivalent to $\sqrt{x^2 - y^2}$ if $x = r\sec(\theta)$ and $y = r\tan(\theta)$ for all real $\theta$?

(A) $-r$
(B) $-|r|$
(C) 1
(D) $-|-r|$
(E) $r$

15. If $r < 0$, which of the following is equivalent to $\sqrt{x^2 + y^2}$ if $x = r\sin(t)$ and $y = r\cos(t)$ for all real values of $t$?

(A) $r$
(B) $-|r|$
(C) $r^2$
(D) $-|-r|$
(E) $-r$

## Answer Key to Practice Problems

1. (C)

2. (E)

3. (C)

4. (E)

5. (B)

6. (C)

7. (B)

8. (C)

9. (A)

10. (C)

11. (E)

12. (B)

13. (D)

14. (E)

15. (E)

## Detailed Solution for Practice Problems

**1.**

According to the piecewise function $f(x)$, $f(2)=3$. Hence, the answer is **(B)**.

**2.**

Let's have a look at the graph.

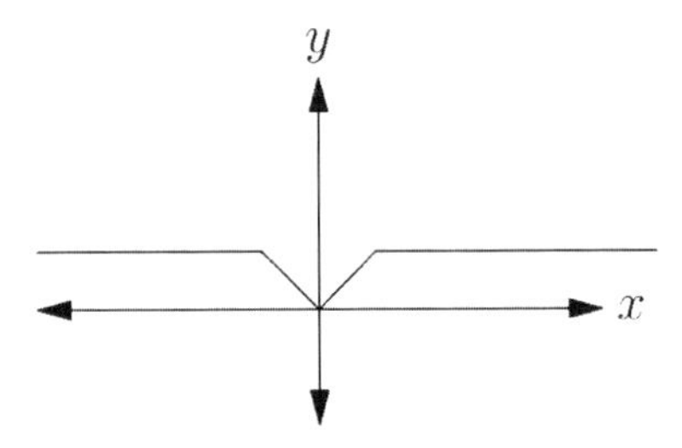


The graph shows that the range is $[0,1]$, so the answer must be **(E)**.

**3.**

In order to find the period from the piecewise function $y=f(x)$, we either locate the distance between the adjacent maxima (or minima). As one can see from the figure below, the period must be 2. Therefore, the answer must be **(C)**.

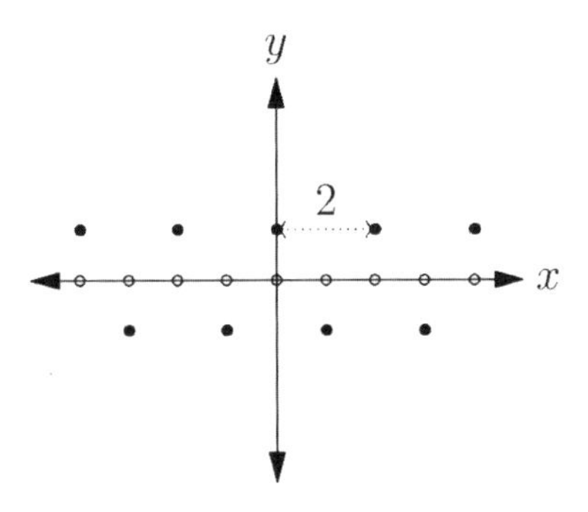


**4.**

Function is continuous if the graph is connected. Therefore, $f(3)=2(3)+b$ must be equal to $f(3)=3(3)-1$. The answer is **(E)**.

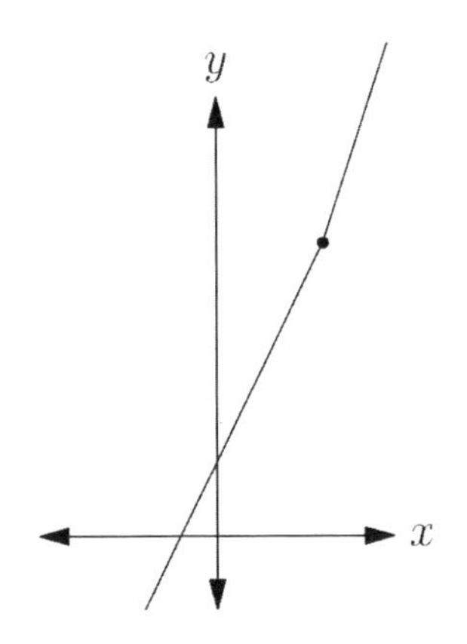

**5.**

$f(-1.46) = \lfloor \lfloor -1.46 \rfloor \rfloor = \lfloor -2 \rfloor = -2$. The answer is **(B)**.

**6.**

$\lfloor -2.45 \rfloor + \lfloor 2.44 \rfloor = -3 + 2 = -1$, so the answer is **(C)**.

**7.**

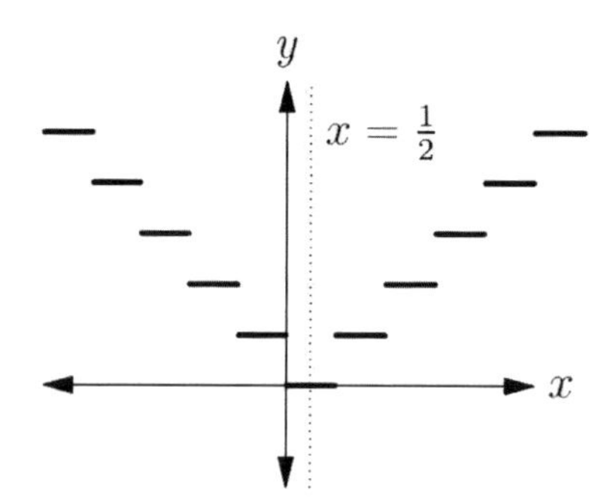


As one can see from the graph of $y = |\lfloor x \rfloor|$, the line of symmetry is at $x = \dfrac{1}{2}$. Hence, the answer is **(B)**.

**8.**

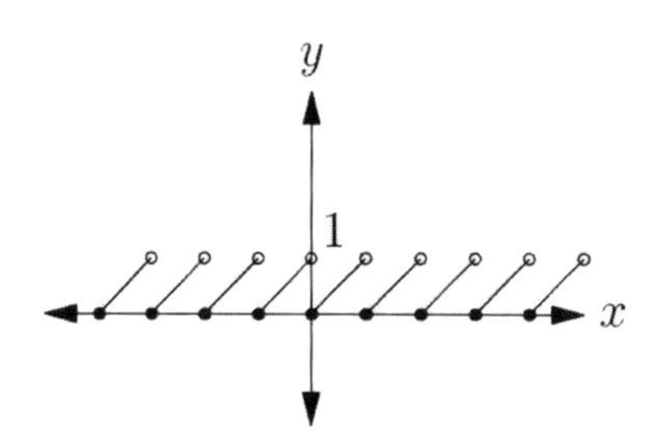


The range of $y = x - \lfloor x \rfloor$ is $[0, 1)$, so the answer is **(C)**.

**9.**

Since $f(x-1) = |4x-3|$, $f(3) = f(4-1) = |4(4)-3| = |13| = 13$, so the answer is **(A)**.

**10.**

Since $|a-b|=|b-a|$, we get $|a-b|-|b-a|=0$. Thus, the answer must be **(C)**.

**11.**

Look at the following graph of $y=\dfrac{|x-2|}{x-2}$.

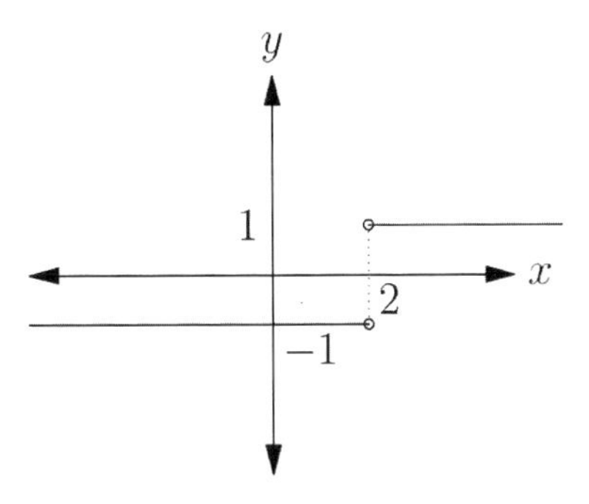


As we approach $x=2$ from the left portion of the graph, $y$-value approaches $-1$. This is called the left-hand limit of $f(x)$ as $x$ approaches 2. On the other hand, as we approach $x=2$ from the right portion of the graph, $y$-value approaches 1. This is called the right-hand limit of $f(x)$ as $x$ approaches 2. Since the left-hand limit and right-hand limit are different, we say $\lim\limits_{x\to 2}\dfrac{|x-2|}{x-2}$ does not exist. Hence, the answer is **(E)**.

**12.**

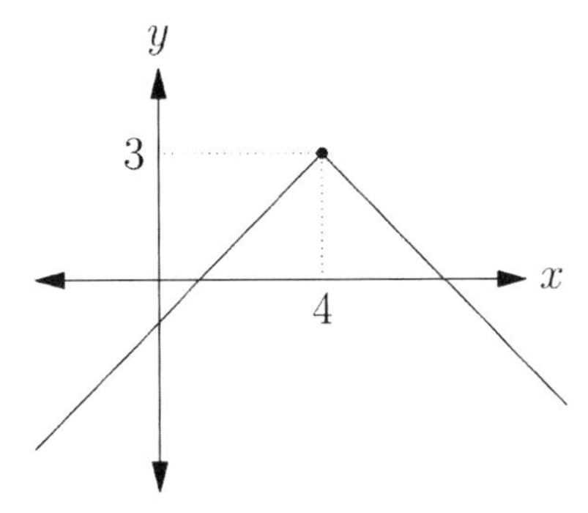


The maximum value of the function $y=-|4-x|+3$ is 3 at $x=4$. Hence, the answer is **(B)**.

**13.**

Substitution is the best strategy to solve this inequality. If we substitute (D), then $|-1|+|2| = 3 < 4$. The answer must be **(D)**. You can easily check from the following graph that $(-1, 2$ is inside the region of $|x|+|y| < 4$.

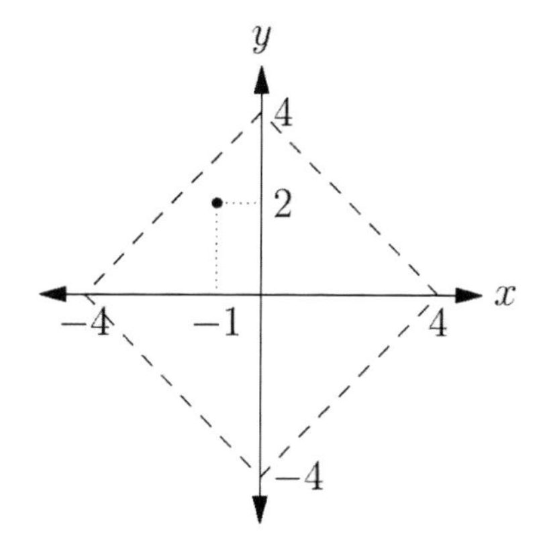


**14.**

$$\begin{aligned}\sqrt{x^2 - y^2} &= \sqrt{(r\sec(\theta))^2 - (r\tan(\theta))^2} \\ &= \sqrt{r^2\sec^2(\theta) - r^2\tan^2(\theta)} \\ &= \sqrt{r^2(\sec^2(\theta) - \tan^2(\theta))} \\ &= \sqrt{r^2} \\ &= |r| \\ &= r\end{aligned}$$

The answer must be **(E)**.

**15.**

$$\begin{aligned}\sqrt{x^2 + y^2} &= \sqrt{(r\sin(t))^2 + (r\cos(t))^2} \\ &= \sqrt{r^2\sin^2(t) + r^2\cos^2(t)} \\ &= \sqrt{r^2(\sin^2(t) + \cos^2(t))} \\ &= \sqrt{r^2} \\ &= |r| \\ &= -r\end{aligned}$$

The correct answer is **(E)**.

# CHECK ON LEARNING #4

## Hyperbolic Trigonometric Identities

$$1+\tan^2(t)=\sec^2(t)$$
$$1+\cot^2(t)=\csc^2(t)$$

First, we begin with $\cos^2(t)+\sin^2(t)=1$. Then, dividing it by $\cos^2(t)$, we get

$$\cos^2(t)+\sin^2(t)=1$$
$$\frac{\cos^2(t)+\sin^2(t)}{\cos^2(t)}=\frac{1}{\cos^2(t)}$$
$$\frac{\cos^2(t)}{\cos^2(t)}+\frac{\sin^2(t)}{\cos^2(t)}=\frac{1}{\cos^2(t)}$$
$$1+\tan^2(t)=\sec^2(t)$$

Second, we divide $\cos^2(t)+\sin^2(t)=1$ by $\sin^2(t)$. Hence,

$$\cos^2(t)+\sin^2(t)=1$$
$$\frac{\cos^2(t)+\sin^2(t)}{\sin^2(t)}=\frac{1}{\sin^2(t)}$$
$$\frac{\cos^2(t)}{\sin^2(t)}+\frac{\sin^2(t)}{\sin^2(t)}=\frac{1}{\sin^2(t)}$$
$$1+\cot^2(t)=\csc^2(t)$$

Always remember that we start from

$$\cos^2(\square)+\sin^2(\square)=1$$

# Topic 7

# Radical Functions and Expressions

✓ Domain and Range of Radical Function

✓ Even and Odd Functions

✓ Radical Expressions and Equations

## 7.1 Domain and Range of Radical Function

The graphs of radical functions are different per index, which means that the domain and range of these functions are different per index.

- If the index is even, then the domain and range are the set of positive real numbers.

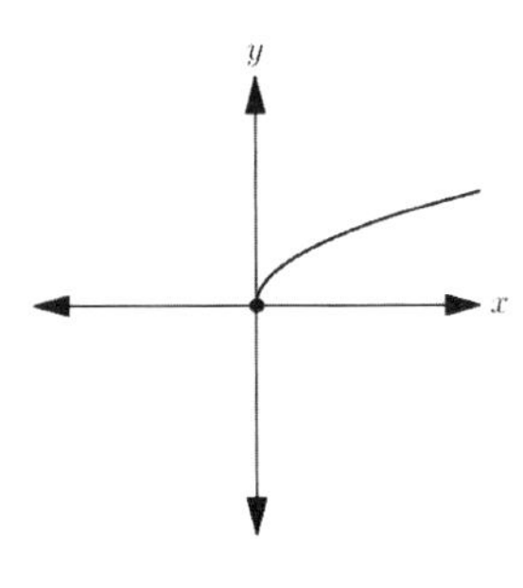


- If the index is odd, then the domain and range are the set of all real numbers.

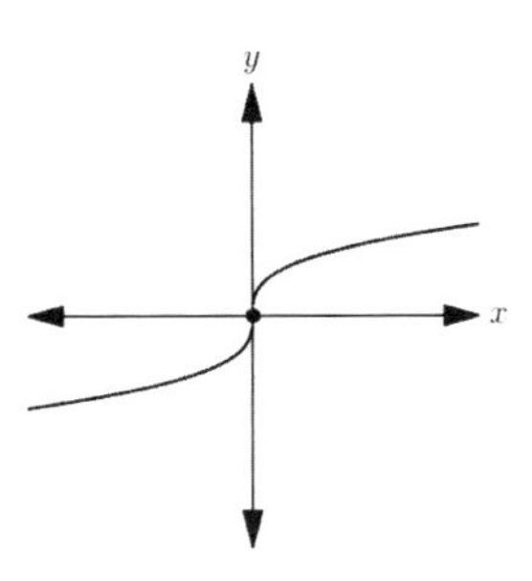


---

**MATHEMATICS LEVEL 2 Test - *Continued***

USE THIS SPACE FOR SCRATCH WORK.

1. Which of the following intervals includes the domain of $f(x) = \sqrt{x-2} + 3$?

(A) $(1, \infty)$
(B) $(2, \infty)$
(C) $(3, \infty)$
(D) $(4, \infty)$
(E) $(5, \infty)$

**MATHEMATICS LEVEL 2 Test - *Continued***

USE THIS SPACE FOR SCRATCH WORK.

2. Which of the following functions has exactly same range as $f(x) = \sqrt{x-2}+3$?

(A) $y = -|2-x|+3$
(B) $y = 4(x-2)+3$
(C) $y = \ln(x-2)+3$
(D) $y = e^{x-2}+3$
(E) $y = (x-2)^2+3$

3. Which of the following functions has different range?

(A) $y = x^3$
(B) $y = \sqrt[3]{x}$
(C) $y = |x|$
(D) $y = \ln(x)$
(E) $y = x$

4. The function $f(x) = \sqrt[4]{|x-1|}+1$ has a value of $x$ for all values of $y$ EXCEPT

(A) 0
(B) 1
(C) 2
(D) 3
(E) 4

## 7.2 Even and Odd Functions

$y = \sqrt[3]{x}$ is an odd function, meaning that the graph is symmetric about the origin. On the other hand, $y = \sqrt{x}$ is neither even nor odd, meaning that the graph is neither symmetric about the $y$-axis nor the origin. Generally speaking,

- we call $f(x)$ even if $f(-x) = f(x)$ for all $x$ in the domain.
- we call $f(x)$ odd if $f(-x) = -f(x)$ for all $x$ in the domain.

Let's have a look at some representatives of even and odd functions.

- Even functions : $y = x^{\text{even integer}}$, $y = \cos(x)$, $y = |x|$, $y = \sec(x)$.
- Odd functions : $y = x^{\text{odd integer}}$, $y = \sin(x)$, $y = \tan(x)$, $y = \csc(x)$, $y = \cot(x)$.

---

**MATHEMATICS LEVEL 2 Test - *Continued***

USE THIS SPACE FOR SCRATCH WORK.

5. Which of the following functions satisfies $f(x) = f(-x)$?

(A) $y = \sqrt[3]{x} + 1$
(B) $y = |x| + 1$
(C) $y = \ln(x) + 1$
(D) $y = \sin(x) + 1$
(E) $y = \tan(x) + 1$

6. Which of the following functions has the graph symmetric about the origin?

(A) $y = 3x^2 + 3$
(B) $y = \sqrt[3]{x} + \sin(x)$
(C) $y = x\tan(x)$
(D) $y = |x| + \cos(x)$
(E) $y = x + x^2$

**MATHEMATICS LEVEL 2 Test - *Continued***

USE THIS SPACE FOR SCRATCH WORK.

7. Which of the following functions satisfies $-f(-x) = f(x)$?

(A) $y = |x|$
(B) $y = \ln(x^2)$
(C) $y = \sin(x^2)$
(D) $y = \cos(x)$
(E) $y = \tan(x)$

8. The graph of $y = \sqrt{|x|}$ is symmetric about

(A) $y = x$
(B) $y = 0$
(C) $x = 0$
(D) $y = -x$
(E) $y = |x|$

9. If $y = f(x)$ is odd function, where $f(1) = 5$, which of the following points must be on its graph?

(A) $(5, 1)$
(B) $(-1, 5)$
(C) $(1, -5)$
(D) $(-1, -5)$
(E) $(-5, -1)$

## 7.3 Simplification of Radical Expressions

Given a radical expression or equation, we should be able to

- rationalize the denominator.
- solve the radical equation.

The conjugate of $a+\sqrt{b}$ is $a-\sqrt{b}$. When we multiply the conjugate to rationalize the denominator, for instance,

$$\frac{1}{a+\sqrt{b}} = \frac{a-\sqrt{b}}{(a+\sqrt{b})(a-\sqrt{b})} = \frac{a-\sqrt{b}}{a^2-b}$$

---

**MATHEMATICS LEVEL 2 Test - *Continued***

USE THIS SPACE FOR SCRATCH WORK.

10. If $a = 2-\sqrt{5}$, then $\left(\frac{1}{a}+2\right)^2 =$

(A) $2+\sqrt{5}$
(B) $2-\sqrt{5}$
(C) 1
(D) 5
(E) $\sqrt{5}$

11. If $\sqrt{6x+1} = \frac{3}{2}$, then $x =$

(A) $\frac{5}{12}$ (B) $\frac{7}{12}$ (C) $\frac{5}{24}$ (D) $\frac{7}{24}$ (E) $\frac{1}{12}$

**MATHEMATICS LEVEL 2 Test - *Continued***

USE THIS SPACE FOR SCRATCH WORK.

12. If $\sqrt[5]{7x+3} = 2$, what is the value of $x$?

(A) $\frac{12}{7}$ (B) $\frac{24}{7}$ (C) 4 (D) $\frac{29}{7}$ (E) 5

13. If $a = \sqrt{7} - \sqrt{5}$, then $(12 - a^2)^2 =$

(A) 35
(B) 70
(C) 140
(D) 280
(E) 560

14. If $x^2 = 11 + 6\sqrt{2}$, then $x =$

(A) $2 + \sqrt{3}$
(B) $2 - \sqrt{3}$
(C) $3 + \sqrt{2}$
(D) $3 - \sqrt{2}$
(E) $-3 + \sqrt{2}$

**MATHEMATICS LEVEL 2 Test - *Continued***

15. Given integers $m$, $n$, where $1 \leq m \leq 10$ and $1 \leq n \leq 36$, for how many ordered pairs of integers $(m, n)$ is $\sqrt{m + \sqrt{n}}$ an integer?

(A) 10
(B) 11
(C) 12
(D) 13
(E) 14

USE THIS SPACE FOR SCRATCH WORK.

## Answer Key to Practice Problems

1. (A)

2. (E)

3. (C)

4. (B)

5. (B)

6. (B)

7. (E)

8. (C)

9. (D)

10. (D)

11. (C)

12. (D)

13. (C)

14. (C)

15. (A)

## Detailed Solution for Practice Problems

**1.**

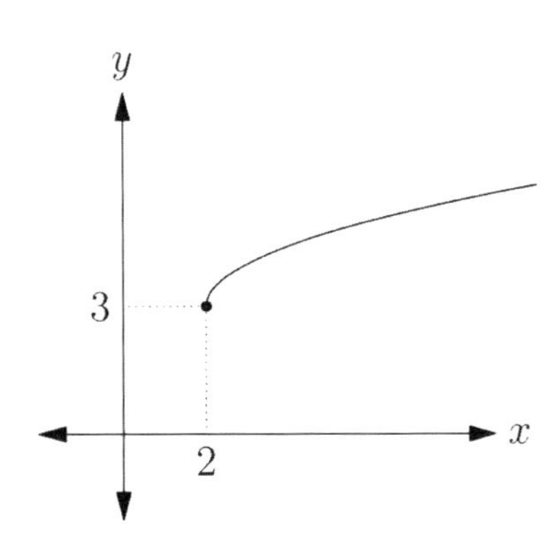


The domain of this function is $[2,\infty)$. The interval that includes the domain is **(A)**.

**2.**

The graph above shows that the range for $y=\sqrt{x-2}+3$ is $[3,\infty)$. The only function that has exactly the same range is $y=(x-2)^2+3$, whose graph is shown below. The answer is **(E)**.

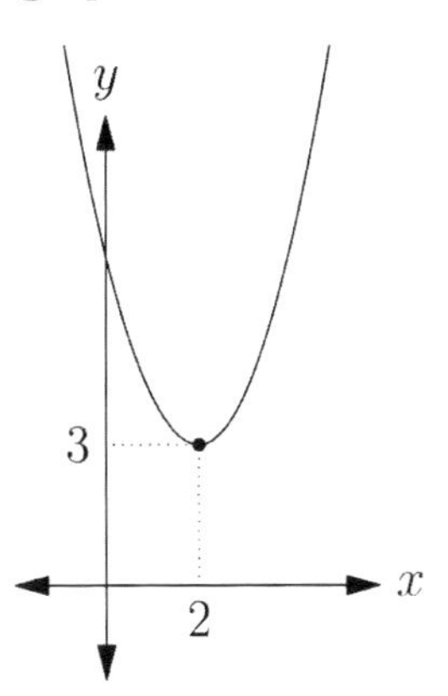


**3.**

- $y=x^3$ has the range of $\mathbb{R}$, the set of real numbers.
- $y=\sqrt[3]{x}$ has the range of $\mathbb{R}$.
- $y=|x|$ has the range of $[0,\infty)$.
- $y=\ln(x)$ has the range of $\mathbb{R}$.
- $y=x$ has the range of $\mathbb{R}$.

Hence, the answer is **(C)**.

**4.**

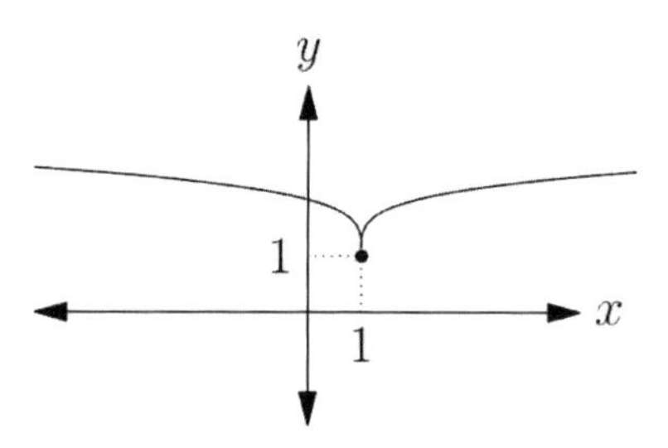


The graph above shows that the range is $[1, \infty)$. Hence, $y = 0$ is not included in its range, so the answer is **(A)**.

**5.**

If $f(x) = f(-x)$, then the function $f(x)$ is even, whose graph is symmetric about the line $x = 0$. The answer must be **(B)**.

**6.**

If the graph is symmetric about the origin, it means the associated function is odd function. The only odd function that matches with this is **(B)**, whose graph can be plotted as the following figure.

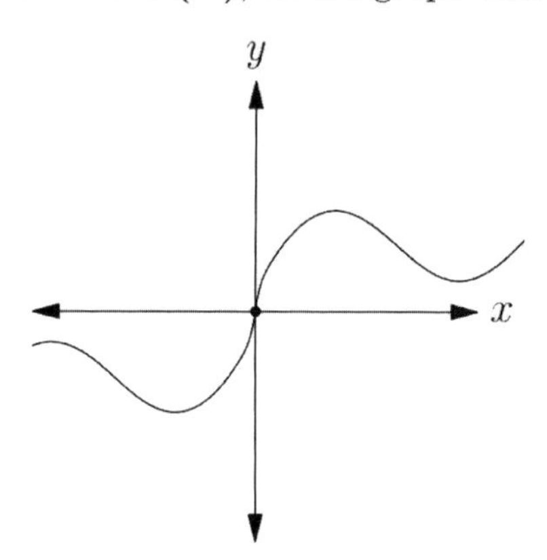


**7.**

Rearranging $-f(-x) = f(x)$, we get $f(-x) = -f(x)$, which is the condition for odd function. The only odd function from the answer choices is **(E)**, i.e., $y = \tan(x)$.

**8.**

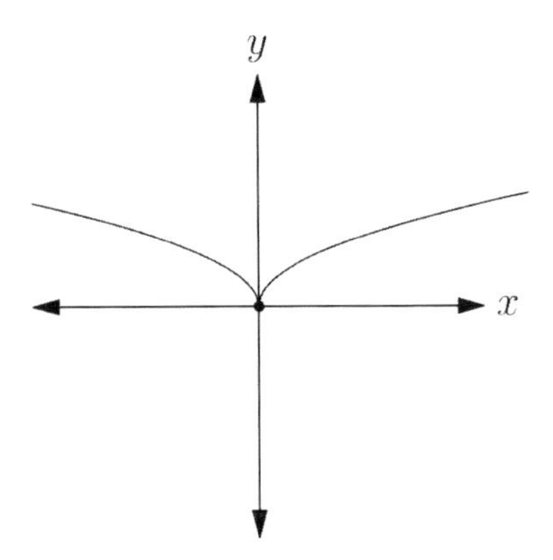


The graph of $y = \sqrt{|x|}$ is symmetric about the $y$-axis, which is $x = 0$, so the answer is **(C)**.

**9.**

If $y = f(x)$ is odd function, then the graph is symmetric about the origin, so $f(1) = 5$ implies that $f(-1) = -5$. Hence, the answer is **(D)**.

**10.**

$$\begin{aligned}\left(\frac{1}{2-\sqrt{5}}+2\right)^2 &= \left(\frac{2+\sqrt{5}}{(2-\sqrt{5})(2+\sqrt{5})}+2\right)^2 \\ &= \left(-2-\sqrt{5}+2\right)^2 \\ &= (-\sqrt{5})^2 \\ &= 5\end{aligned}$$

The answer is **(D)**.

**11.**

$$\begin{aligned}\sqrt{6x+1} &= \frac{3}{2} \\ 6x+1 &= \frac{9}{4} \\ 6x &= \frac{5}{4} \\ x &= \frac{5}{24}\end{aligned}$$

The answer must be **(C)**.

**12.**

$$\begin{aligned}\sqrt[5]{7x+3} &= 2\\ 7x+3 &= 2^5\\ 7x+3 &= 32\\ 7x &= 29\\ x &= \frac{29}{7}\end{aligned}$$

The answer must be **(D)**.

**13.**

Let $a=\sqrt{7}-\sqrt{5}$. Then,

$$\begin{aligned}(12-a^2)^2 &= (12-(\sqrt{7}-\sqrt{5})^2))^2\\ &= (12-(7-2\sqrt{35}+5))^2\\ &= (12-(12-2\sqrt{35}))^2\\ &= (2\sqrt{35})^2\\ &= 4\cdot 35\\ &= 140\end{aligned}$$

Hence, the answer is **(C)**.

**14.**

Let's use the answer choices.

- $(2+\sqrt{3})^2 = 4+4\sqrt{3}+3 = 7+4\sqrt{3} \neq 11+6\sqrt{2}$.
- $(2-\sqrt{3})^2 = 4-4\sqrt{3}+3 = 7-4\sqrt{3} \neq 11+6\sqrt{2}$.
- $(3+\sqrt{2})^2 = 9+6\sqrt{2}+2 = 11+6\sqrt{2}$.
- $(3-\sqrt{2})^2 = 9-6\sqrt{2}+2 = 11-6\sqrt{2} \neq 11+6\sqrt{2}$.
- $(-3+\sqrt{2})^2 = 9-6\sqrt{2}+2 = 11-6\sqrt{2} \neq 11+6\sqrt{2}$.

The answer must be **(C)**.

**15.**

In order for $\sqrt{m+\sqrt{n}}$ to be an integer, we first have to make $\sqrt{n}$ as an integer, which means that $n$ should be a perfect square.

Case 1. $n=1$. In this case, $\sqrt{m+\sqrt{n}}=\sqrt{m+\sqrt{1}}=\sqrt{m+1}$ has to be an integer. Therefore, $m+1$ has to be a perfect square, where $2 \leq m+1 \leq 11$. There are only two perfect squares between 2 and 11, i.e., $m+1=4,9$. Hence, $m=3$ or 8, i.e., $(m,n)=(3,1),(8,1)$.

Case 2. $n=4$. In this case, $\sqrt{m+\sqrt{n}}=\sqrt{m+\sqrt{4}}=\sqrt{m+2}$ must be a perfect square, where $3 \leq m+2 \leq 12$. Just like in case 1, there are two perfect squares between 3 and 12, i.e., $m+2=4,9$. Hence, $(m,n)=(2,4),(7,4)$.

Case 3. $n=9$. In this case, $\sqrt{m+\sqrt{n}}=\sqrt{m+\sqrt{9}}=\sqrt{m+3}$ must be a perfect square where $4 \leq m+3 \leq 13$. There are two perfect squares between 4 and 13, inclusive i.e., $m+3=4,9$, so $(m,n)=(1,9),(5,9)$.

Case 4. $n=16$. In this case, $\sqrt{m+\sqrt{n}}=\sqrt{m+\sqrt{16}}=\sqrt{m+4}$ must be a perfect square where $5 \leq m+4 \leq 14$. There is only one perfect square between 5 and 14, i.e., $m+4=9$, so $(m,n)=(5,16)$.

Case 5. $n=25$. In this case, $\sqrt{m+\sqrt{n}}=\sqrt{m+\sqrt{25}}=\sqrt{m+5}$ must be a perfect square where $6 \leq m+5 \leq 15$. There is only one perfect square between 6 and 15, i.e., $m+5=9$, so $(m,n)=(4,25)$.

Case 6. $n=36$. In this case, $\sqrt{m+\sqrt{n}}=\sqrt{m+\sqrt{36}}=\sqrt{m+6}$ must be a perfect square where $7 \leq m+6 \leq 16$. There are two perfect squares between 7 and 16, inclusive, i.e., $m+6=9,16$, so $(m,n)=(3,36),(10,36)$.

To sum up, there are $10(=2+2+2+1+1+2)$ pairs of $m$ and $n$ that make $\sqrt{m+\sqrt{n}}$ an integer. Hence, the answer must be **(A)**.

# Topic 8

# Rational Functions and Expressions

- ✓ Undefined values of $x$ and $y$
- ✓ Limit as $x$ approaches $\infty$
- ✓ Rational expressions with other functions

## 8.1 Undefined values of $x$ and $y$

Typically, when a rational function is given, we are given with

$$f(x) = \underbrace{\frac{p(x)}{q(x)}}_{\text{Original}} = \underbrace{\frac{r(x)}{t(x)}}_{\text{Simplified}}$$

and the question asks us to find the domain, vertical asymptotes or holes.

- All values of $x$ such that $q(x) = 0$ are undefined.
- If we simplify, the zeros of the canceled term(s) are holes.
- The zeros of $t(x)$ are vertical asymptotes.

Sometimes, the question asks us about the undefined value(s) of $y$. In this case, such undefined value(s) of $y$ are horizontal asymptote(s). In order to solve it algebraically, rewrite $y = f(x)$ into $x = g(y)$, and look at the denominators of $g(y)$. For instance,

$$y = \underbrace{\frac{x-2}{3+x}}_{f(x)\text{ part}} \Longrightarrow y(3+x) = x-2 \Longrightarrow x(y-1) = -2-3y \Longrightarrow x = \underbrace{-\frac{2+3y}{y-1}}_{g(y)\text{ part}}$$

where $y \neq 1$ in the last part.

---

**MATHEMATICS LEVEL 2 Test - *Continued***

1. If $f(x) = \dfrac{1-2x}{(2x-1)(x^2+1)}$, for which value of $s$ is $f(x)$ undefined?

(A) $-\frac{1}{2}$ (B) $-1$ (C) $0$ (D) $1$ (E) $\frac{1}{2}$

USE THIS SPACE FOR SCRATCH WORK.

**MATHEMATICS LEVEL 2 Test - *Continued***

USE THIS SPACE FOR SCRATCH WORK.

2. If $f(x) = \dfrac{(2-x)(x+1)}{2x-4}$, what is the value of $x$ such that the graph of $f(x)$ is discontinuous?

(A) $-1$ (B) $-2$ (C) 0 (D) 2 (E) 1

3. Which of the following $y$ is undefined for $y = \dfrac{x-1}{x+1}$?

(A) $-\dfrac{1}{2}$ (B) $-1$ (C) 0 (D) 1 (E) $\dfrac{1}{2}$

4. Which value of $k$ does $f(x) = \dfrac{4x+12}{x+k}$ have no vertical asymptote?

(A) 4
(B) 3
(C) 12
(D) $-3$
(E) $-4$

## 8.2 Limit as $x$ approaches $\infty$

You will find some questions related to limit process in the exam, whenever you find the phrase

as SOMETHING **approaches** SOMETHING

If you find this phrase, please use the limit expression in your calculator. Also, they might ask you to find the **horizontal asymptote** of a rational function, which is another application of limit process.

- Degrees are equal : $y = \dfrac{3x^1+1}{x^1+2}$ has a horizontal asymptote of $y = \dfrac{3}{1}$, which is a ratio of leading coefficients.
- Denominator degree is greater : $y = \dfrac{3x^1+1}{x^2+1}$ has a horizontal asymptote of $y = 0$, WIHTOUT EXCEPTION.
- Denominator degree is smaller : $y = \dfrac{x^2+1}{3x^1+1}$ has NO horizontal asymptote, WITHOUT EXCEPTION.

---

**MATHEMATICS LEVEL 2 Test** - ***Continued***

USE THIS SPACE FOR SCRATCH WORK.

5. What value does $\dfrac{x^2-1}{x+1}$ approach as $x$ approaches 1?

(A) $\dfrac{1}{2}$ (B) $-1$ (C) 0 (D) 1 (E) $\dfrac{1}{2}$

6. The function $f(x) = \dfrac{x^2-9}{x-3}$ is undefined for $x =$

(A) $-3$
(B) $-9$
(C) 1
(D) 9
(E) 3

**MATHEMATICS LEVEL 2 Test - *Continued***

USE THIS SPACE FOR SCRATCH WORK.

7. What value does $y = \frac{2+10x^2}{3-4x}$ approach as $x$ gets infinitely large?

(A) $\frac{1}{\sqrt{5}}$ (B) $-\infty$ (C) $\frac{3}{4}$ (D) $\infty$ (E) $-\frac{1}{\sqrt{5}}$

8. Which of the following has different horizontal asymptote?

(A) $y = \frac{2x+1}{x-3}$

(B) $y = \frac{4x+1}{2x-1}$

(C) $y = \frac{1-2x}{4-x}$

(D) $y = \frac{1-2x}{3+4x}$

(E) $y = \frac{2x^2+1}{x^2+3}$

9. $\lim_{x \to 1} \frac{x^2-3x+2}{x-1} =$

(A) 1

(B) $-1$

(C) 0

(D) 2

(E) $-2$

## MATHEMATICS LEVEL 2 Test - *Continued*

USE THIS SPACE FOR SCRATCH WORK.

10. Which of the following functions has a different horizontal asymptote?

(A) $y = \frac{2x-1}{4x+3}$

(B) $y = \frac{x^2}{3+2x^2}$

(C) $y = \frac{3-5x}{7-10x}$

(D) $y = \frac{2-4x}{3+2x}$

(E) $y = \frac{1+x^2}{2x^2-5}$

11. What value does $y = \frac{2e^x-1}{3e^x+1}$ approach as $x$ gets infinitely small?

(A) $\frac{2}{3}$ (B) 1 (C) $-1$ (D) $\frac{3}{2}$ (E) $-\frac{2}{3}$

12. What is the range of $y = f(x) = \frac{|x-1|}{x-1}$?

(A) the set of all real numbers

(B) the set of all positive real numbers

(C) the set of real numbers except $x = 1$

(D) $\{1\}$

(E) $\{1, -1\}$

## 8.3 Rational expressions with other functions

In SAT Math Level 2, it is likely to have a rational function with other functions in either denominator of numerator. Typical questions related to rational expressions are domain/range questions. We simply have to find the **commonly valid** intervals of $x$ or $y$ for such functions.

- Make denominator non-zero.
- Canceled terms make hole(s).
- Numerator contains $x$-intercepts.

Also, reciprocal function frequently appears in the test. Given $y = f(x)$, then $y = \dfrac{1}{f(x)}$ can be graphed using the following rules.

- Sign does not change.
- Increasing portion turns into decreasing portion, vice versa.
- $x$-intercepts turn into vertical asymptotes, vice versa.

---

**MATHEMATICS LEVEL 2 Test - *Continued***

USE THIS SPACE FOR SCRATCH WORK.

13. The domain of $f(x) = \dfrac{\sqrt{3-x}}{x-1}$ does not include the value of $x =$

(A) 3
(B) 2
(C) 1
(D) 0
(E) $-1$

## MATHEMATICS LEVEL 2 Test - *Continued*

USE THIS SPACE FOR SCRATCH WORK.

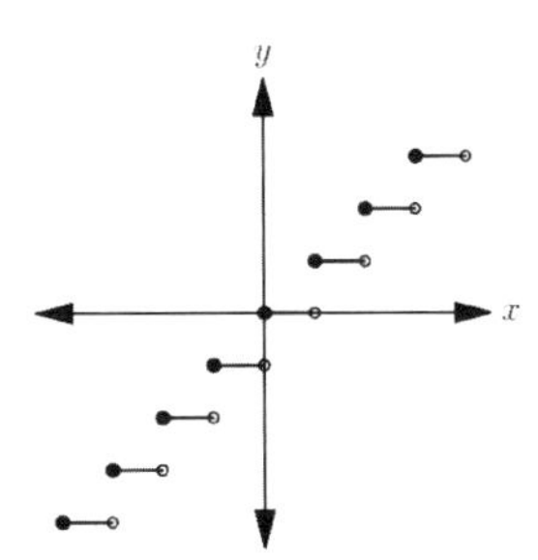


Figure. $y = \lfloor x \rfloor$

14. Which of the following values of $y$ is undefined for $y = \dfrac{1}{\lfloor x \rfloor}$?

(A) $-1$ (B) $\dfrac{1}{2}$ (C) 1 (D) $-\dfrac{1}{2}$ (E) 2

15. The function $f(x) = \dfrac{x}{\lfloor x \rfloor} - 1$ has its $x$-intercepts at

(A) the set of all integers
(B) the set of non-zero integers
(C) the set of all real numbers
(D) the set of positive integers
(E) the set of negative integers

## Answer Key to Practice Problems

1. (E)

2. (D)

3. (D)

4. (B)

5. (C)

6. (E)

7. (B)

8. (D)

9. (B)

10. (D)

11. (C)

12. (E)

13. (C)

14. (E)

15. (B)

## Detailed Solution for Practice Problems

**1.**

Given $f(x) = \dfrac{1-2x}{(2x-1)(x^2+1)}$, the function is undefined for $x = \dfrac{1}{2}$ because there is a hole at $x = \dfrac{1}{2}$. The answer is **(E)**.

**2.**

Given $f(x) = \dfrac{(2-x)(x+1)}{2(x-2)}$, the function is undefined, due to a hole, at $x = 2$. Hence, the function $y = f(x)$ is discontinuous at $x = 2$. The answer must be **(D)**.

**3.**

$$\begin{aligned} y &= \frac{x-1}{x+1} \\ y(x+1) &= x-1 \\ yx+y &= x-1 \\ y+1 &= x-yx \\ \frac{y+1}{1-y} &= x \end{aligned}$$

There is no $x$-value for $y = 1$. Hence, $y = 1$ is undefined for $y = \dfrac{x-1}{x+1}$. The answer is **(D)**.

**4.**

If $k = 3$, then $f(x) = \dfrac{4x+12}{x+k} = \dfrac{4(x+3)}{x+3} = 4$ where $x \neq -3$. Hence, the function $f(x)$ has a hole at $x = -3$, for $k = 3$. Hence, the answer is **(B)**.

**5.**

$$\begin{aligned} \lim_{x\to 1} \frac{x^2-1}{x+1} &= \lim_{x\to 1} \frac{(x-1)(x+1)}{x+1} \\ &= \lim_{x\to 1}(x-1) \\ &= 0 \end{aligned}$$

The answer is **(C)**.

**6.**

$f(x) = \dfrac{x^2-9}{x-3} = \dfrac{(x-3)(x+3)}{x-3} = x+3$ is undefined for $x=3$. The answer is **(E)**.

**7.**

$$\begin{aligned}\lim_{x\to\infty}\frac{2+10x^2}{3-4x} &= \lim_{x\to\infty}\frac{2/x+10x}{3/x-4}\\ &= \frac{0+\infty}{0-4}\\ &= -\infty\end{aligned}$$

Hence, the answer is **(B)**.

**8.**

- $\displaystyle\lim_{x\to\pm\infty}\frac{2x+1}{x-3} = 2.$
- $\displaystyle\lim_{x\to\pm\infty}\frac{4x+1}{2x-1} = 2.$
- $\displaystyle\lim_{x\to\pm\infty}\frac{1-2x}{4-x} = 2.$
- $\displaystyle\lim_{x\to\pm\infty}\frac{1-2x}{3+4x} = -\frac{1}{2}.$
- $\displaystyle\lim_{x\to\pm\infty}\frac{2x^2+1}{x^2+3} = 2.$

The answer must be **(D)**.

**9.**

$$\begin{aligned}\lim_{x\to 1}\frac{x^2-3x+2}{x-1} &= \lim_{x\to 1}\frac{(x-2)(x-1)}{x-1}\\ &= \lim_{x\to 1}(x-2)\\ &= (1-2)\\ &= -1\end{aligned}$$

The answer is **(B)**.

**10.**

- $\lim_{x\to\pm\infty} \dfrac{2x-1}{4x+3} = \dfrac{2}{4} = \dfrac{1}{2}.$
- $\lim_{x\to\pm\infty} \dfrac{x^2}{3+2x^2} = \dfrac{1}{2}.$
- $\lim_{x\to\pm\infty} \dfrac{3-5x}{7-10x} = \dfrac{-5}{-10} = \dfrac{1}{2}.$
- $\lim_{x\to\pm\infty} \dfrac{2-4x}{3+2x} = \dfrac{-4}{2} = -2.$
- $\lim_{x\to\pm\infty} \dfrac{1+x^2}{2x^2-5} = \dfrac{1}{2}.$

The answer must be **(D)**.

**11.**

$$\begin{aligned}\lim_{x\to-\infty} \frac{2e^x-1}{3e^x+1} &= \frac{2(0)-1}{3(0)+1} \\ &= \frac{-1}{1} \\ &= -1\end{aligned}$$

Hence, the answer is **(C)**.

**12.**

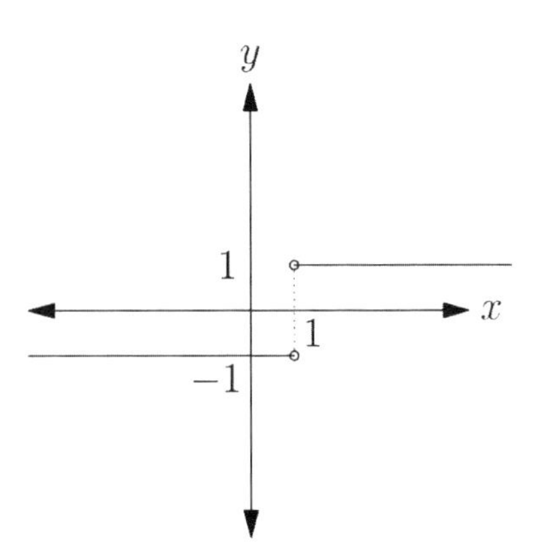


The set of $y$-values is $\{-1, 1\}$, so the answer must be **(E)**.

## 13.

The domain must satisfy the following two conditions.

- $3 - x \geq 0$, so $3 \geq x$.
- $x - 1 \neq 0$, so $x \neq 1$.

Hence, the answer is **(C)**.

## 14.

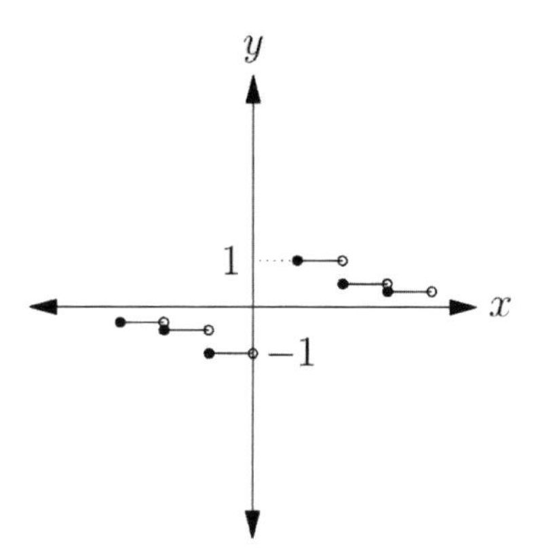


As one can check from the graph of $y = \dfrac{1}{\lfloor x \rfloor}$, the range consists of $\dfrac{1}{k}$ form where $k$ is a nonzero integer. Hence, the answer must be **(E)**.

## 15.

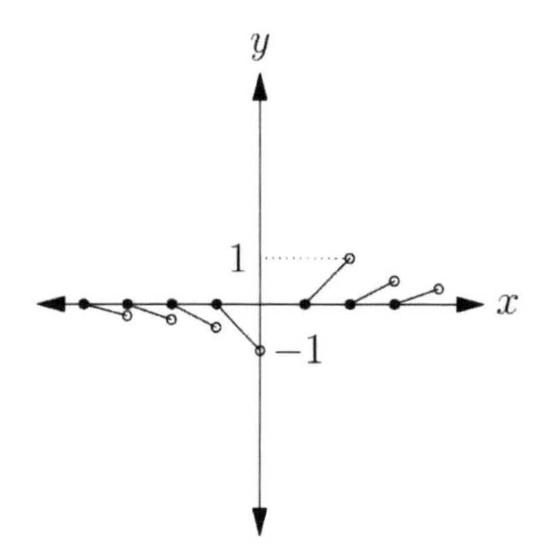


As one can see from the graph of $f(x) = \dfrac{x}{\lfloor x \rfloor} - 1$, the $x$-intercepts are located at all non-zero integers. Therefore, the answer is **(B)**.

# CHECK ON LEARNING #5

## Exponential Limits

$$\lim_{x \to \infty} \frac{\mathbf{A}e^x + B}{\mathbf{C}e^x + D} = \frac{A}{C}$$

$$\lim_{x \to -\infty} \frac{Ae^x + \mathbf{B}}{Ce^x + \mathbf{D}} = \frac{B}{D}$$

As $x$ grows, the limit value approaches $\frac{\infty}{\infty}$, which is an indeterminate form. We choose the largest term of the denominator to divide both numerator and denominator term.

$$\begin{aligned}\lim_{x \to \infty} \frac{\mathbf{A}e^x + B}{\mathbf{C}e^x + D} &= \lim_{x \to \infty} \frac{A + B/e^x}{C + D/e^x} \\ &= \frac{A+0}{C+0} \\ &= \frac{A}{C}\end{aligned}$$

On the other hand, as $x$ grows smaller, the limit value DOES NOT approach $\frac{0}{0}$. We only divide both numerator and denominator by the largest term of the denominator if the expression is in indeterminate form. In other words,

$$\begin{aligned}\lim_{x \to \infty} \frac{Ae^x + \mathbf{B}}{Ce^x + \mathbf{D}} &= \frac{A \times 0 + B}{C \times 0 + D} \\ &= \frac{B}{D}\end{aligned}$$

Topic 9

# Trigonometric Functions and Identities

- ✓ Trigonometric Ratio
- ✓ Trigonometric Function
- ✓ Trigonometric Identities
- ✓ Laws of Cosines and Laws of Sines
- ✓ Vectors and Complex Numbers

## 9.1 Trigonometric Ratio

Given a right triangle,

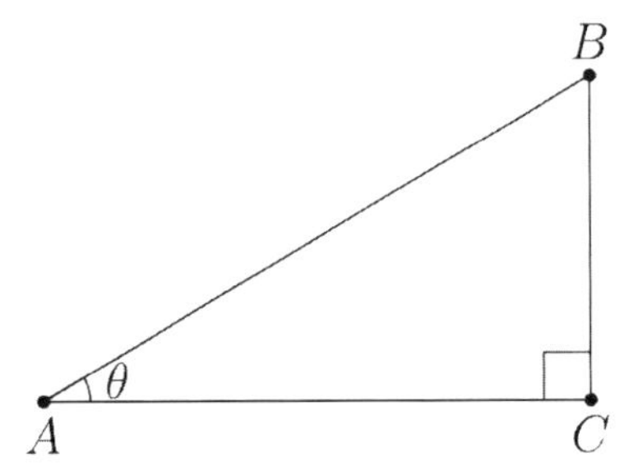


- $\sin(\theta) = \dfrac{\text{Opposite}}{\text{Hypotenuse}}$ and $\csc(\theta) = \dfrac{1}{\sin(\theta)}$
- $\cos(\theta) = \dfrac{\text{Adjacent}}{\text{Hypotenuse}}$ and $\sec(\theta) = \dfrac{1}{\cos(\theta)}$
- $\tan(\theta) = \dfrac{\text{Opposite}}{\text{Adjacent}}$ and $\cot(\theta) = \dfrac{1}{\tan(\theta)}$

Typical SAT Math Level 2 questions ask us to change one ratio to another one. Some difficult questions involve signs, which can be elaborated by C.A.S.T.

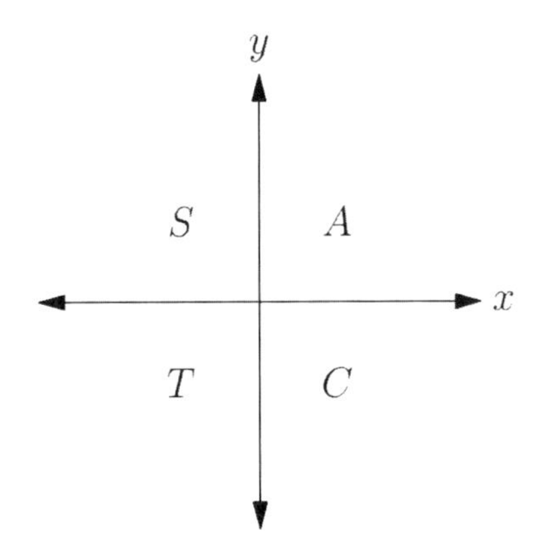


Assuming that we rotate the positive $x$-axis counterclockwise[1],

- All angles formed in the quadrant I are positive.
- Sine angles formed in the quadrant II are positive.
- Tangent angles formed in quadrant III are positive.
- Cosine angles formed in quadrant IV are positive.

[1]If we rotate the positive $x$-axis(i.e. a ray) clockwise, we consider such angle as negative angle.

**MATHEMATICS LEVEL 2 Test - *Continued***

USE THIS SPACE FOR SCRATCH WORK.

1. If $\cot(x) = 5$, then $\dfrac{\tan(x)}{\cot(x)} =$

(A) $\frac{1}{25}$ (B) 5 (C) $\frac{1}{5}$ (D) 1 (E) 25

2. The cosine of an angle is one third the sine of the same angle. What is the tangent of the angle?

(A) 3 (B) 1 (C) $\frac{1}{3}$ (D) $-1$ (E) $-\frac{1}{3}$

3. If $\tan(t) = \frac{x}{2}$ for $0 < t < \frac{\pi}{2}$, then $\cos(t) =$

(A) $\dfrac{2}{\sqrt{x^2+4}}$

(B) $\dfrac{2}{\sqrt{x^2-4}}$

(C) $\dfrac{x}{\sqrt{x^2+4}}$

(D) $\dfrac{x}{\sqrt{x^2-4}}$

(E) $\dfrac{x}{\sqrt{4-x^2}}$

**MATHEMATICS LEVEL 2 Test - *Continued***

USE THIS SPACE FOR SCRATCH WORK.

4. If $\sin(x) = m$ for $90° < x < 180°$, then $\tan(x) =$

(A) $\frac{1}{m^2}$

(B) $\frac{m}{\sqrt{1-m^2}}$

(C) $\frac{1-m^2}{m}$

(D) $-\frac{m}{1-m^2}$

(E) $-\frac{m}{\sqrt{1-m^2}}$

5. If $\cos(x) = n$ for $180° < x < 270°$, then $\cot(x) =$

(A) $\frac{1}{n^2-1}$

(B) $\frac{n}{\sqrt{1-n^2}}$

(C) $\frac{1-n^2}{n}$

(D) $-\frac{n}{1-n^2}$

(E) $-\frac{n}{\sqrt{1-n^2}}$

## 9.2 Trigonometric Function

First off, let's have a look at sinusoidal (both sine and cosine) functions.

$$f(x) = \underbrace{A}_{|A|=\text{amplitude}} \times \cos\left(\underbrace{B}_{\text{Period } = \frac{2\pi}{|B|}}\left(x \pm \underbrace{h}_{\text{Horizontal shift}}\right)\right) \pm \underbrace{k}_{\text{Vertical shift}}$$

- The graph of $y = \sin(x)$ is bounded between $y = 1$ and $y = -1$, inclusive, whose one period is given by the following diagram.

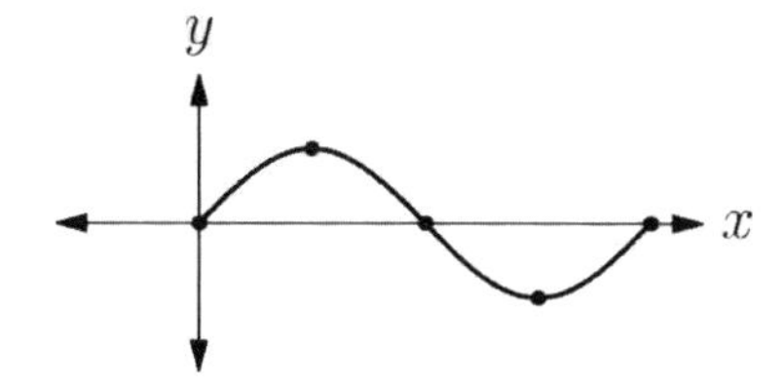


- The graph of $y = \cos(x)$ is also bounded between $y = 1$ and $y = -1$, inclusive, whose one period is given by the following diagram, different from that of $y = \sin(x)$.

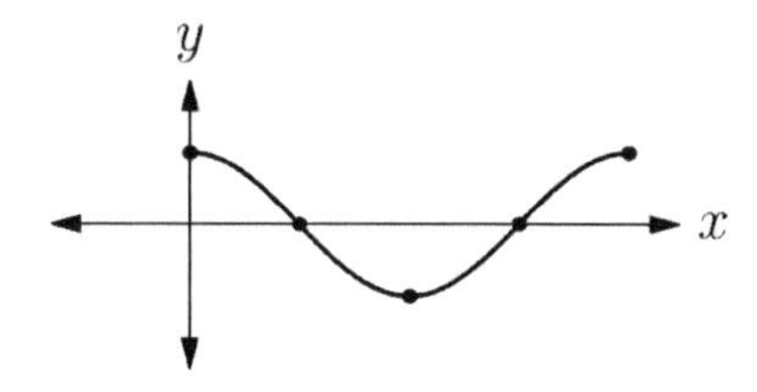


---

**MATHEMATICS LEVEL 2 Test - *Continued***

6. What is the value of an acute angle $x$ when $2\cos(x) - \sqrt{2} = 0$?

USE THIS SPACE FOR SCRATCH WORK.

(A) 30°

(B) 60°

(C) 45°

(D) 25°

(E) 27.5°

## MATHEMATICS LEVEL 2 Test - *Continued*

USE THIS SPACE FOR SCRATCH WORK.

7. If $\sin(x)\cos(x) < 0$, then which of the following must be true?

I. $\tan(x) < 0$

II. $\csc(x)\cot(x) < 0$

III. The primary angle $x$ is in the second or fourth quadrant.

(A) I only
(B) II only
(C) III only
(D) II and III only
(E) I and III only

8. For $0° \leq x \leq 45°$, what is the least possible value of $1+\cos(x)$?

(A) 1.09
(B) 1.47
(C) 1.71
(D) 1.91
(E) 1.24

9. What is the period of $y = 2\cos(\pi - 2\pi x) + 1$?

(A) 2
(B) 1
(C) $-1$
(D) $-2$
(E) None of the above

Talking about the period, we should be able to solve problems related to periodic functions.

- The original period of $y = \cos(x)$, $y = \sin(x)$, $y = \sec(x)$ and $y = \csc(x)$ is $2\pi$.
- The original period of $y = \tan(x)$ and $y = \cot(x)$ is $\pi$.
- The original period of $y = |$ any trigonometric function $|$ is $\pi$.

Especially, we should be able to understand the algebraic expression that shows periodic property of a given function.

$$f(x) = f(\underbrace{x \pm k}_{k \text{ is the period}})$$

In other words, if $f(x) = f(x+2)$, then $f(-1) = f(1) = f(3) = \cdots$. It's easy to process this as a number jumps between parantheses with the difference of period, either adding the period or subtracting it from the original number.

Also, amplitude is not simply the absolute value of the coefficient in front of $\cos(x)$ or $\sin(x)$, but half the difference between maximum and minimum.

$$\text{Amplitude} = \frac{\text{maximum} - \text{minimum}}{2}$$

---

**MATHEMATICS LEVEL 2 Test - *Continued***

10. What is the period of $y = 1 - 3|\cos(\pi x - \frac{\pi}{2})|$?

USE THIS SPACE FOR SCRATCH WORK.

(A) 3
(B) 2
(C) 1
(D) −1
(E) 2

**MATHEMATICS LEVEL 2 Test - *Continued***

USE THIS SPACE FOR SCRATCH WORK.

11. If $f(x)$ is periodic with the period of 3, which of the following is equal to $f(11)$?

(A) $f(7)$
(B) $f(5)$
(C) $f(3)$
(D) $f(1)$
(E) $f(0)$

12. If the water level of a leaking tank is modeled by

$$H(t) = 124.321 + 4.1516\cos(\frac{\pi}{2}(t-3))$$

for $t$ minutes after the initial fill-up, in which it has a faucet turned on to fill the leakage, what is the difference between the maximum and minimum water level for the first 10 minutes?

(A) 4.1516
(B) 8.3032
(C) −4.1516
(D) 124.321
(E) 128.4726

## 9.3 Trigonometric Identities

Trigonometric identities are something that saves you time during the exam. Also, some questions cannot be solved without using the identities. Here are the most common types of identities that have appeared in the exam.

- $\cos^2(x)+\sin^2(x)=1$
- $\sec^2(x)-\tan^2(x)=1$
- $\cos(x)=\sin(y) \Longleftrightarrow x+y=\frac{\pi}{2}$
- $\sin(2x)=2\sin(x)\cos(x)$
- $\cos(2x)=\cos^2(x)-\sin^2(x)=2\cos^2(x)-1=1-2\sin^2(x)$
- $\sin(x)=\sin(y) \Longleftrightarrow x+y=\cdots,-\pi,\pi,3\pi,5\pi,\cdots$
- $\cos(x)=\cos(y) \Longleftrightarrow x+y=\cdots,-2\pi,0,2\pi,4\pi,\cdots$

---

**MATHEMATICS LEVEL 2 Test - *Continued***

USE THIS SPACE FOR SCRATCH WORK.

13. For $z<0$, if $a=z\cos\theta$ and $b=z\sin\theta$, then $\sqrt{a^2+b^2}=$

(A) $-z$
(B) $z$
(C) $2z$
(D) $z\cos\theta\sin\theta$
(E) $z(\cos\theta+\sin\theta)$

**MATHEMATICS LEVEL 2 Test - *Continued***

USE THIS SPACE FOR SCRATCH WORK.

14. What is the amplitude of $y = 5 + 6\cos^2(x)$?

(A) 2
(B) 3
(C) 4
(D) 6
(E) 12

15. $\sin(\frac{\pi}{2} + t) =$

(A) $\sin(\frac{\pi}{3} - t)$
(B) $\sin(2t)$
(C) $\sin(t)$
(D) $\cos(2t)$
(E) $\cos(t)$

16. $\cos(\frac{\pi}{2} - \theta) =$

(A) $\sin(\theta)$
(B) $\cos(\theta)$
(C) $\sin(\theta)\cos(\theta)$
(D) $\sin(\frac{\pi}{2} - \theta)$
(E) $\sin(2\theta)$

**MATHEMATICS LEVEL 2 Test - *Continued***

USE THIS SPACE FOR SCRATCH WORK.

17. The maximum value of $4\sin(x)\cos(x)$ is equal to

(A) $\frac{1}{2}$ (B) 1 (C) 2 (D) 4 (E) $\frac{3\sqrt{3}}{2}$

18. If $x+y=90°$, which of the following must be true?

(A) $\cos(x)=\cos(y)$
(B) $\sin(x)=-\sin(y)$
(C) $\tan(x)=\cot(y)$
(D) $\sin(x)+\cos(y)=1$
(E) $\tan(x)+\cot(y)=1$

19. If $f(x)=3x^2$ and $g(x)=f(\sin(x))+f(\cos(x))$, then $g(35°)$ is equal to

(A) 1
(B) 3
(C) 3.15
(D) 3.96
(E) 4

## 9.4 Laws of Cosines and Sines

The laws of cosines and sines are used to find out the measure of length or angle of a triangle.

- Given $\triangle ABC$, $\frac{\sin(A)}{a} = \frac{\sin(B)}{b} = \frac{\sin(C)}{c}$.
- Given $\triangle ABC$, the area is equal to $\frac{1}{2}ab\sin(C) = \frac{1}{2}bc\sin(A) = \frac{1}{2}ca\sin(B)$.
- Given $\triangle ABC$,

$$\begin{aligned} a^2 &= b^2 + c^2 - 2bc\cos(A) \\ b^2 &= a^2 + c^2 - 2ac\cos(B) \\ c^2 &= a^2 + b^2 - 2ab\cos(C) \end{aligned}$$

Using both laws of sines and cosines, we eventually retrieve Heron's formula to find out the area of a given triangle.

$$\text{Area} = \sqrt{s(s-a)(s-b)(s-c)} \text{ where } s = \frac{a+b+c}{2}$$

---

**MATHEMATICS LEVEL 2 Test - *Continued***

USE THIS SPACE FOR SCRATCH WORK.

20. If $m\angle A = 100°$, $m\angle B = 50°$, and $AB = 5$, what is the length of $\overline{BC}$?

(A) 1.736
(B) 5.685
(C) 9.848
(D) 12.660
(E) 14.396

21. If a triangle has sides of length 5, 6, and 7, what is the cosine of the largest interior angle in the triangle?

(A) 0 (B) $\frac{2}{3}$ (C) $\frac{1}{5}$ (D) $\frac{3}{4}$ (E) $\frac{1}{3}$

**MATHEMATICS LEVEL 2 Test - *Continued***

USE THIS SPACE FOR SCRATCH WORK.

22. What is the area of the triangle $ABC$ if $AB = 3$, $AC = 5$ and $m\angle BAC = 40°$?

(A) 1.928
(B) 3.214
(C) 4.821
(D) 5.745
(E) 9.642

23. A triangle has sides measuring $4, 5, 6$. What is the secant of the largest interior angle?

(A) $\frac{1}{8}$ (B) $-8$ (C) 4 (D) 8 (E) $-\frac{1}{8}$

24. What is the area of equilateral triangle whose side length is 10?

(A) 10.7
(B) 12.5
(C) 25.1
(D) 43.3
(E) 86.6

**MATHEMATICS LEVEL 2 Test - *Continued***

USE THIS SPACE FOR SCRATCH WORK.

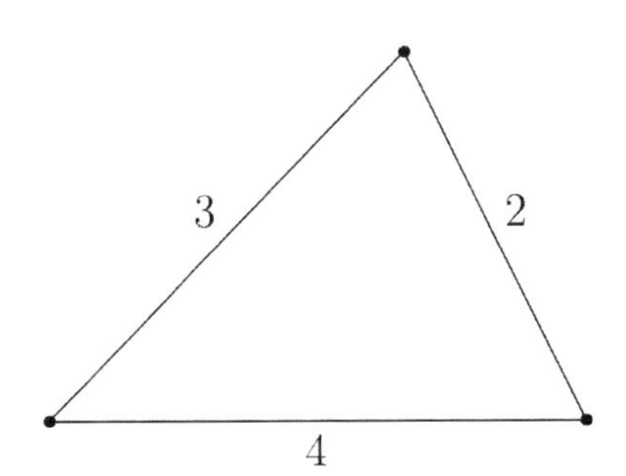


25. The area of the triangle $ABC$, rounded to the nearest tenth, is equal to

(A) 1.8
(B) 2.5
(C) 2.9
(D) 3.2
(E) 3.5

26. How many non-congruent triangles are there if $AB = 20$, $m\angle A = 60^\circ$, and $BC = 15$?

(A) 0
(B) 1
(C) 2
(D) 3
(E) 4

## 9.5 Vectors and Complex Numbers

Vector is an object with both magnitude and direction. If we write it $< a,b >$, then it represents an arrow that starts from $(0,0)$ and ends at $(a,b)$.

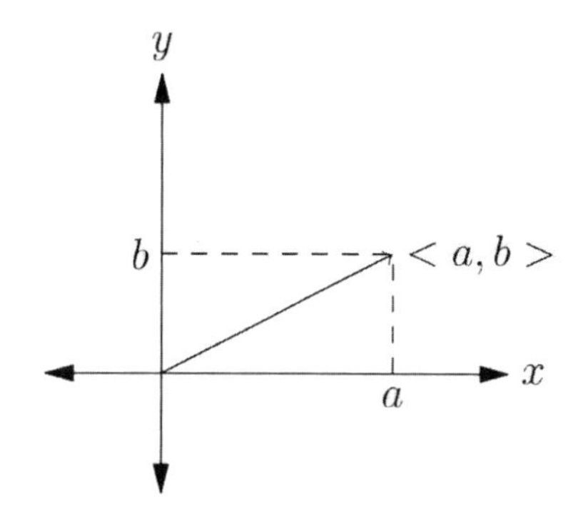


If the length of the vector is $r = \sqrt{a^2+b^2}$, and the angle $\theta$ formed by the vector and the $x$-axis satisfies $\tan(\theta) = \frac{b}{a}$, then

$$< a,b > = < r\cos(\theta), r\sin(\theta) >$$

Vector questions are directly connected to the laws of cosines, i.e.,

$$(\mathbf{r}+\mathbf{s})^2 = \mathbf{r}^2 + \mathbf{s}^2 - 2\times\mathbf{r}\times\mathbf{s}\times\cos(\alpha)$$

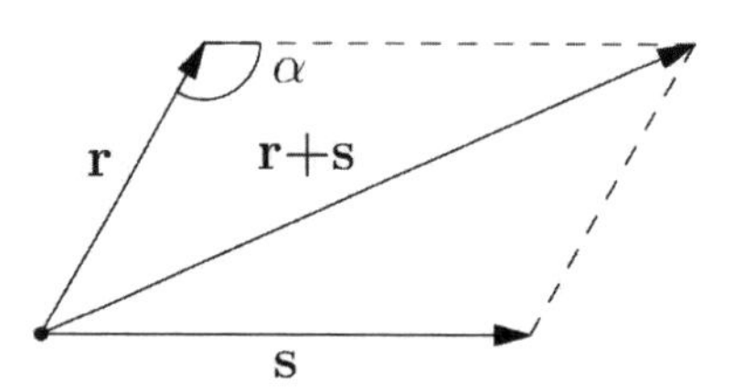


**MATHEMATICS LEVEL 2 Test - *Continued***

27. If there are two vectors $\mathbf{r}$ and $\mathbf{s}$ where the angle formed between $\mathbf{r}$ and $\mathbf{s}$ is $30°$ and $|\mathbf{r}| = 3$ and $|\mathbf{s}| = 4$, then the length of $\mathbf{r}+\mathbf{s}$?

(A) 2.05
(B) 2.88
(C) 5.12
(D) 6.77
(E) 9.84

USE THIS SPACE FOR SCRATCH WORK.

## MATHEMATICS LEVEL 2 Test - *Continued*

Questions 27-28 use the same diagram below.

USE THIS SPACE FOR SCRATCH WORK.

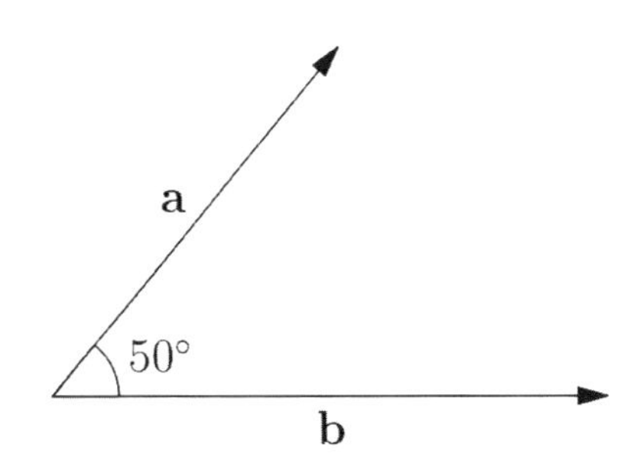


28. If two vectors **a** and **b** have the length of 4 and 5, what is the length of $\mathbf{a}-\mathbf{b}$?

(A) 2.45
(B) 2.76
(C) 2.85
(D) 3.91
(E) 8.18

29. If two vectors **a** and **b** have the length of 4 and 5, what is the length of $2\mathbf{a}+3\mathbf{b}$?

(A) 17.5
(B) 18.7
(C) 21.1
(D) 22.4
(E) 24.9

Complex numbers are written in terms of real part and imaginary part, where

$$z = a + bi$$

where $a$ is the real part of $z$, denoted by $\mathrm{Re}(z)$, and $b$ the imaginary part of $z$, denoted by $\mathrm{Im}(z)$. In the complex plane, the usual $y$-axis is called the imaginary axis, the $x$-axis the real axis, as shown in the figure below.

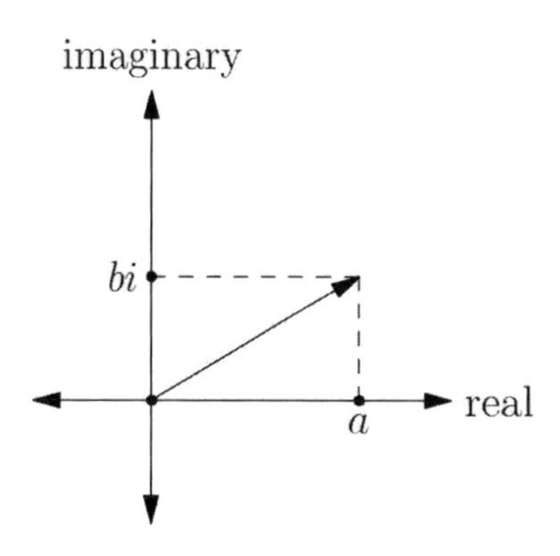


The reason why this section is part of Trigonometry is because we express complex numbers in trigonometric form, i.e.,

$$z = \sqrt{a^2 + b^2}(\cos(\theta) + i\sin(\theta))$$

where $\theta$ is the angle formed by the positive $x$-axis and the vector $< a, b >$.

---

**MATHEMATICS LEVEL 2 Test - *Continued***

USE THIS SPACE FOR SCRATCH WORK.

30. If $z = 3 + 4i$ and $w = 2 - 3i$, what is the value of $2z + 3w$?

(A) $2 + 3i$
(B) $12 - i$
(C) $12 + i$
(D) $6 - 12i$
(E) $6 + 12i$

## Answer Key to Practice Problems

1. (A)
2. (A)
3. (A)
4. (E)
5. (E)
6. (C)
7. (E)
8. (C)
9. (B)
10. (C)
11. (B)
12. (B)
13. (A)
14. (B)
15. (E)
16. (A)
17. (C)
18. (C)
19. (B)
20. (C)
21. (C)
22. (C)
23. (D)
24. (D)
25. (C)
26. (C)
27. (D)
28. (D)
29. (C)
30. (B)

## Detailed Solution for Practice Problems

**1.**

Since $\cot(x) = 5$, then $\tan(x) = \frac{1}{5}$. Therefore, $\frac{\tan(x)}{\cot(x)} = \frac{1/5}{5} = \frac{1}{25}$. The answer must be **(A)**.

**2.**

Let $\theta$ be the angle of our interest. Then, $\cos(\theta) = \frac{1}{3}\sin(\theta)$. Hence, $\tan(\theta) = \frac{\sin(\theta)}{\cos(\theta)} = 3$. The answer is **(A)**.

**3.**

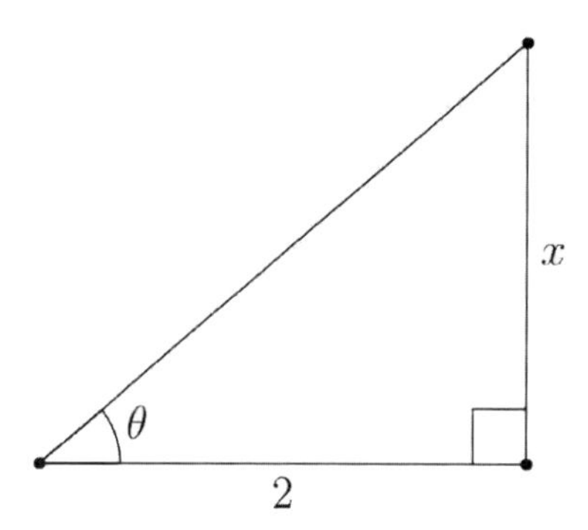


If $\tan(t) = \frac{x}{2}$, then the figure above shows that the length of hypotenuse is $\sqrt{x^2+4}$. Hence, $\cos(t) = \frac{2}{\sqrt{x^2+4}}$, so the answer is **(A)**.

**4.**

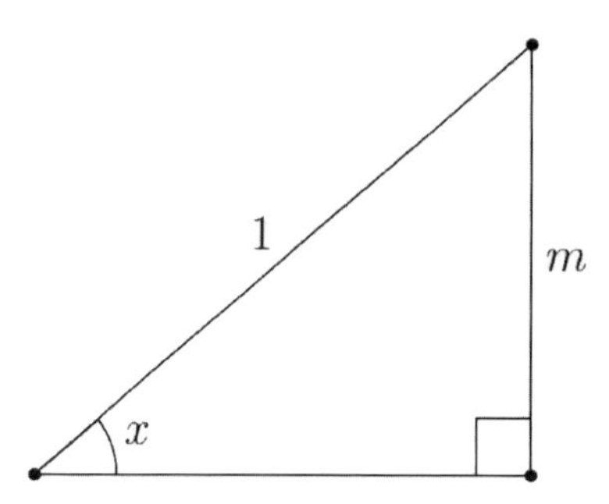


The adjacent length of a right triangle is $\sqrt{1-m^2}$, so $|\tan(x)| = \frac{m}{\sqrt{1-m^2}}$. Since the angle is obtuse, according to C.A.S.T, $\tan(x) < 0$. Therefore, $\tan(x) = -\frac{m}{\sqrt{1-m^2}}$, so the answer is **(E)**.

**5.**

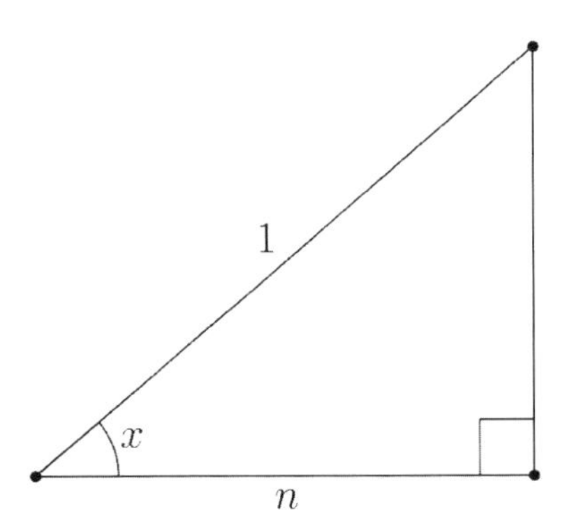


As one can check from the figure above, the opposite leg's length must be $\sqrt{1-n^2}$, by Pythagorean theorem. Now, $|\tan(x)| = \dfrac{\sqrt{1-n^2}}{n}$, so $|\cot(x)| = \dfrac{n}{\sqrt{1-n^2}}$. Since $180^\circ < x < 270^\circ$ implies that the angle is in the third quadrant, it means that $\cos(x) = n < 0$. So, $\cot(x) = -\dfrac{n}{\sqrt{1-n^2}}$ because $-n > 0$. Therefore, the answer must be **(E)**.

**6.**

$$\begin{aligned} 2\cos(x) - \sqrt{2} &= 0 \\ 2\cos(x) &= \sqrt{2} \\ \cos(x) &= \frac{\sqrt{2}}{2} \\ x &= \cos^{-1}\left(\frac{\sqrt{2}}{2}\right) \\ x &= 45^\circ \end{aligned}$$

Hence, the answer is **(C)**.

**7.**

If $\sin(x)\cos(x) < 0$, then $(\sin(x), \cos(x)) = (+,-), (-,+)$. This means that the primary angle rotated from the positive $x$-axis in counterclockwise direction must stay either in the second or fourth quadrants.

- $\tan(x) < 0$ means $x$ is in the second or fourth quadrants.
- $\csc(x)\cot(x) < 0$ means that $\dfrac{\cos(x)}{\sin^2(x)} < 0$, so $\cos(x) < 0$. This means that the angle $x$ stays in the second or third quadrants.
- This is just a paraphrase of $\sin(x)\cos(x) < 0$.

Thus, the answer is **(E)**.

**8.**

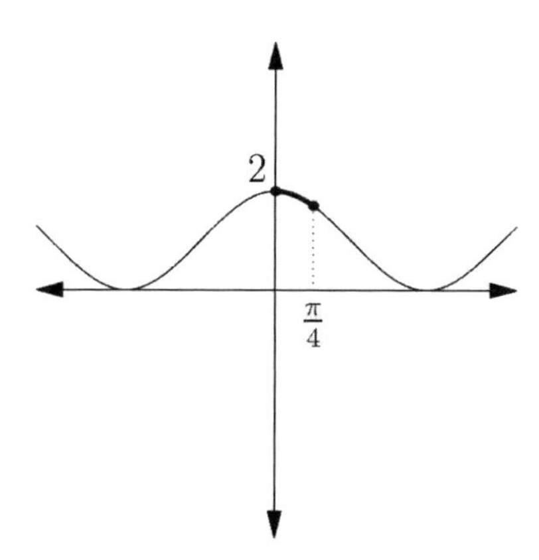


The least possible value of $1+\cos(x)$ occurs at $x = 45°$(or $x = \frac{\pi}{4}$ radian). The answer is **(C)**.

**9.**

$y = 2\cos(-2\pi(x-\frac{1}{2}))+1$ has the period of $\frac{2\pi}{|-2\pi|} = 1$, so the answer is **(B)**.

**10.**

$y = 1-3|\cos(\pi(x-\frac{1}{2}))$ has the period of $\frac{\pi}{\pi} = 1$, so the answer is **(C)**.

**11.**

If $f(x)$ is periodic with the period of 3, then $f(11) = f(8) = f(5) = f(2) = f(-1) = \cdots$. Hence, the answer is **(B)**.

**12.**

Instead of graphing the function, we can directly calculate the difference between maximum and minimum values by multiplying 2 to the amplitude. The answer is **(B)**, i.e., $2(4.1516) = 8.3032$.

**13.**

$$\begin{aligned}
\sqrt{a^2+b^2} &= \sqrt{z^2\cos^2\theta + z^2\sin^2\theta} \\
&= \sqrt{z^2(\cos^2(\theta)+\sin^2(\theta))} \\
&= \sqrt{z^2} \\
&= |z| \\
&= -z
\end{aligned}$$

The answer is **(A)**.

**14.**

$$\begin{aligned} y &= 5+6\cos^2(x) \\ &= 5+6\left(\frac{1+\cos(2x)}{2}\right) \\ &= 5+3(1+\cos(2x)) \\ &= 8+3\cos(2x) \end{aligned}$$

The amplitude of $y=5+6\cos^2(x)$ is 3, so the answer is **(B)**.

**15.**

Since $\sin(\frac{\pi}{2}+t)=\sin(\frac{\pi}{2}-t)=\cos(t)$, the answer is **(E)**.

**16.**

$\cos(\frac{\pi}{2}-\theta)=\sin(\theta)$, so the answer is **(A)**.

**17.**

$4\sin(x)\cos(x)=2(2\sin(x)\cos(x))=2\sin(2x)$, and $-2\leq 2\sin(2x)\leq 2$ implies that the maximum value is 2. Therefore, the answer is **(C)**.

**18.**

If two angles $x$ and $y$ are complementary, then

- $\cos(x)=\sin(y)$
- $\tan(x)=\cot(y)$

Hence, the answer is **(C)**.

**19.**

$$\begin{aligned} g(35^\circ) &= f(\sin(35^\circ))+f(\cos(35^\circ)) \\ &= 3\sin^2(35^\circ)+3\cos^2(35^\circ) \\ &= 3(\sin^2(35^\circ)+\cos^2(35^\circ)) \\ &= 3 \end{aligned}$$

Therefore, the answer is **(B)**.

**20.**

According to the laws of sines,

$$\begin{aligned}\frac{BC}{AB} &= \frac{\sin(100^\circ)}{\sin(30^\circ)}\\ BC &= AB \times \frac{\sin(100^\circ)}{\sin(30^\circ)}\\ BC &\approx 9.848\end{aligned}$$

Hence, the answer is **(C)**.

**21.**

Let the largest angle be $\theta$. Then, according to the laws of cosines,

$$\begin{aligned}7^2 &= 5^2 + 6^2 - 2(5)(6)\cos(\theta)\\ 49 &= 25 + 36 - 60\cos(\theta)\\ -12 &= -60\cos(\theta)\\ \frac{1}{5} &= \cos(\theta)\end{aligned}$$

Hence, the answer must be **(C)**.

**22.**

According to the laws of sines, the area of triangle $ABC$, denoted by $[ABC]$, equals

$$\begin{aligned}[ABC] &= \frac{1}{2} \times 3 \times 5 \times \sin(40^\circ)\\ &\approx 4.821\end{aligned}$$

The answer is **(C)**.

**23.**

Let the largest interior angle be $\theta$. According to the laws of cosines,

$$\begin{aligned}6^2 &= 4^2 + 5^2 - 2(4)(5)\cos(\theta)\\ 36 &= 16 + 25 - 40\cos(\theta)\\ -5 &= -40\cos(\theta)\\ \frac{1}{8} &= \cos(\theta)\\ 8 &= \sec(\theta)\end{aligned}$$

Thus, the answer is **(D)**.

**24.**

Given an equilateral triangle with side length of 10, the area of a triangle equals

$$\begin{aligned}\frac{1}{2}\times 10\times 10\times \sin(60^\circ) &= 50\times\frac{\sqrt{3}}{2}\\ &= 25\sqrt{3}\\ &\approx 43.3\end{aligned}$$

The answer is **(D)**.

**25.**

Using Heron's formula, we will find $s=\dfrac{2+3+4}{2}=4.5$, the length of semi-perimeter, first. Then, according to Heron's formula, the area is given by

$$\begin{aligned}\text{Area} &= \sqrt{s(s-2)(s-3)(s-4)}\\ &= \sqrt{4.5(4.5-2)(4.5-3)(4.5-4)}\\ &= \sqrt{4.5(2.5)(1.5)(0.5)}\\ &\approx 2.9\end{aligned}$$

The answer is **(C)**.

**26.**

By SSA ambiguous case, since $BC > AB\sin(60^\circ)$, there are two non-congruent triangles $ABC$, so the answer is **(C)**. For more information, if $BC = AB\sin(60^\circ)$, then there should be one right triangle formed. On the other hand, if $BC < AB\sin(60^\circ)$, there is no triangle formed whatsoever.

**27.**

$$\begin{aligned}|\mathbf{r}+\mathbf{s}|^2 &= 3^2+4^2-2(3)(4)\cos(150^\circ)\\ &= 9+16-24\cos(150^\circ)\\ &= 25-24\cos(150^\circ)\\ &\approx 45.784\end{aligned}$$

Therefore, $|\mathbf{r}+\mathbf{s}|\approx\sqrt{45.784}\approx 6.766\approx 6.77$, so the answer is **(D)**.

**28.**

$$\begin{aligned}|\mathbf{a}-\mathbf{b}|^2 &= 4^2+5^2-2(4)(5)\cos(50^\circ)\\ &= 16+25-40\cos(50^\circ)\\ &= 41-40\cos(50^\circ)\\ &\approx 15.29\end{aligned}$$

Hence, $|\mathbf{a}-\mathbf{b}| \approx \sqrt{15.29} \approx 3.91$, so the answer is **(D)**.

**29.**

$$\begin{aligned}|2\mathbf{a}+3\mathbf{b}|^2 &= (2(4))^2+(3(5))^2-2(8)(15)\cos(130^\circ)\\ &= 8^2+15^2-240\cos(130^\circ)\\ &= 289-240\cos(130^\circ)\\ &\approx 443.27\end{aligned}$$

Hence, $|2\mathbf{a}+3\mathbf{b}| \approx \sqrt{443.27} \approx 21.1$, so the answer is **(C)**.

**30.**

$$\begin{aligned}2z+3w &= 2(3+4i)+3(2-3i)\\ &= (6+8i)+(6-9i)\\ &= (6+6)+(8-9)i\\ &= 12-i\end{aligned}$$

Thus, the answer must be **(B)**.

# CHECK ON LEARNING #6

## Law of Cosines

$$a^2 = b^2 + c^2 - 2bc\cos(A)$$
$$b^2 = a^2 + c^2 - 2ac\cos(B)$$
$$c^2 = a^2 + b^2 - 2ab\cos(C)$$

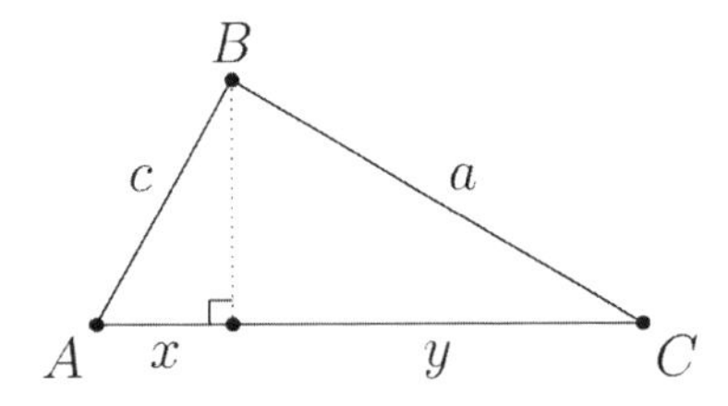


As one can see from the figure above, we can deduce the first law of cosines, i.e.,

$$b = c\cos(A) + a\cos(C)$$
$$a = b\cos(C) + c\cos(B)$$
$$c = a\cos(B) + b\cos(A)$$

Then, the second law of cosines can be deduced by producing square terms.

$$b^2 = bc\cos(A) + ab\cos(C)$$
$$a^2 = ab\cos(C) + ac\cos(B)$$
$$c^2 = ac\cos(B) + bc\cos(A)$$

Hence, we add or subtract three equations to retrieve the law of cosines. To be specific,

$$a^2 - b^2 - c^2 = -2bc\cos(A)$$
$$b^2 - c^2 - a^2 = -2ac\cos(B)$$
$$c^2 - a^2 - b^2 = -2ab\cos(C)$$

# Topic 10

# Miscellaneous Functions

- ✓ Periodic Function
- ✓ Recursive Function with Transformation
- ✓ Parametric Function

## 10.1 Periodic Function

Trigonometric function is a major example of a periodic function. In the subject test, however, there are some questions related to the algebraic property of general periodic function.

First, refresh periodic functions related to trigonometry. Remember the following lists of information.

- Sinusoidal function(including cosecant or secant functions) has a period of $2\pi$.
- Tangent or cotangent function has a period of $\pi$.

General formula to remember is given by

$$f(x) = f(x+k) \text{ where } k \text{ is } \textbf{period}.$$

General formula for the periodic function oftentimes appear in the subject test, and it is quite useful to know this concept beforehand.

**MATHEMATICS LEVEL 2 Test - *Continued***

USE THIS SPACE FOR SCRATCH WORK.

1. If $f(x-4) = f(x)$, and $f(1) = 2$ and $f(2) = 3$, what is the value of $f(-3) + f(-2)$?

(A) 2 (B) 3 (C) 4 (D) 5 (E) 6

2. What is the period of $|\cot(2x)|$?

(A) $2\pi$ (B) $\frac{3}{2}\pi$ (C) $\pi$ (D) $\frac{\pi}{2}$ (E) $\frac{\pi}{4}$

## 10.2 Recursive Function

Recursive functions are written in the form of $f(x+k) = mf(x)$. In fact, this is simply a transformation of a function. In the subject test, functions written recursively frequently appear in many forms, and they can be easily solved by **value-substitution**. Let's have a look at how a problem can be solved by the value-substitution.

---

**MATHEMATICS LEVEL 2 Test - *Continued***

USE THIS SPACE FOR SCRATCH WORK.

3. If $f(x+2) = 3f(x)$ and $f(1) = 3$, what is the value of $f(5)$?

(A) 1 (B) 3 (C) 9 (D) 27 (E) 81

4. If $f(x-2) = 2f(x)$ and $f(x) = 2$ for $1 \leq x < 3$, what is true about $f(x)$ for $3 \leq x < 5$?

(A) $f(x) = 4$
(B) $f(x) = 2$
(C) $f(x) = 1$
(D) $f(x) = 0.5$
(E) $f(x) = 0.25$

## 10.3 Parametric Functions

In the subject test, the time-parameter $t$ is used for parametric functions. So far, most of the functions we deal with connect $x$ and $y$ directly. By adding $t$-variable, a point $(x,y)$ can be traced by different $t$-values.

First, instead of expressing relation between $x$ and $y$, time parameter becomes an independent variable to express $x$ and $y$, respectively, i.e.,

$$x = x(t)$$
$$y = y(t)$$

Second, **polar coordinates** show up in the test. A rectangular coordinates $(x,y)$ can be transformed into $(r,\theta)$. Think about polar coordinates as a Navy submarine whose navigation screen is all green, where a straight-line is rotated counter-clockwise. Here, $r$ represents the radius, and $\theta$ the angle formed by counter-clockwise rotation from the $x$-axis. The following formulas are helpful when we solve polar coordinates questions.

- $(x,y) = (r\cos(\theta), r\sin(\theta))$
- $r^2 = x^2 + y^2$ and $\tan(\theta) = \frac{y}{x}$.

---

**MATHEMATICS LEVEL 2 Test - *Continued***

5. Which of the following is NOT equivalent to the polar coordinates $(1, \frac{\pi}{3})$?

USE THIS SPACE FOR SCRATCH WORK.

(A) $(1, \frac{7\pi}{3})$
(B) $(-1, -\frac{2\pi}{3})$
(C) $(-1, \frac{4\pi}{3})$
(D) $(1, -\frac{5\pi}{3})$
(E) $(1, -\frac{7\pi}{3})$

**MATHEMATICS LEVEL 2 Test - *Continued***

USE THIS SPACE FOR SCRATCH WORK.

6. A line has parametric equations $x(t) = 8 - t$ and $y(t) = 10 + 2t$ where $t$ is the parameter. What is the slope of the line?

(A) $-1$

(B) $-2$

(C) 26

(D) 2

(E) $\frac{1}{2}$

7. What is the graph of the parametric equations $x(t) = \sec(t)$ and $y(t) = \tan(t)$?

(A) Parabola
(B) Circle
(C) Hyperbola
(D) Ellipse
(E) Line

8. Given a parametric equation $x(t) = 6 - 2t$ and $y(t) = t - 4$, where $t$ is the parameter, what is the $x$-intercept?

(A) 4
(B) 0
(C) $-2$
(D) 2
(E) $-4$

**MATHEMATICS LEVEL 2 Test - *Continued***

USE THIS SPACE FOR SCRATCH WORK.

9. Given two points defined by parametric equations such that $(x(t), y(t)) = (4t - 3, 0)$ and $(a(t), b(t)) = (0, 2 - 4t)$, at which of the following $t$-values would they be closest to each other?

(A) 0.875
(B) 0.625
(C) 0.375
(D) 0.125
(E) 0

10. Which of the following quadratic curves does $r = \dfrac{1}{1 + \cos(\theta)}$ represent?

(A) a line
(B) a parabola
(C) a hyperbola
(D) a circle
(E) a point

## Answer Key to Practice Problems

1. (D)

2. (D)

3. (D)

4. (A)

5. (E)

6. (B)

7. (C)

8. (C)

9. (B)

10. (B)

## Detailed Solution for Practice Problems

1.

First of all, if $f(x-4)=f(x)$, it means that the function is periodic with the period of 4. Therefore, $f(-3)=f(1)=f(5)=\cdots$, and $f(-2)=f(2)=f(6)=\cdots$. Hence,
$f(-3)+f(-2)=f(1)+f(2)=2+3=5$, so the answer is **(D)**.

2.

Absolute-valued trigonometric functions always have the original period of $\pi$. Therefore, $|\cot(2x)|$ has the period of $\frac{\pi}{2}$, so the answer is **(D)**.

3.

$$\begin{aligned} f(3) &= 3f(1) \\ &= 3(3) \\ &= 9 \\ f(5) &= 3f(3) \\ &= 3(9) \\ &= 27 \end{aligned}$$

The answer must be **(D)**.

4.

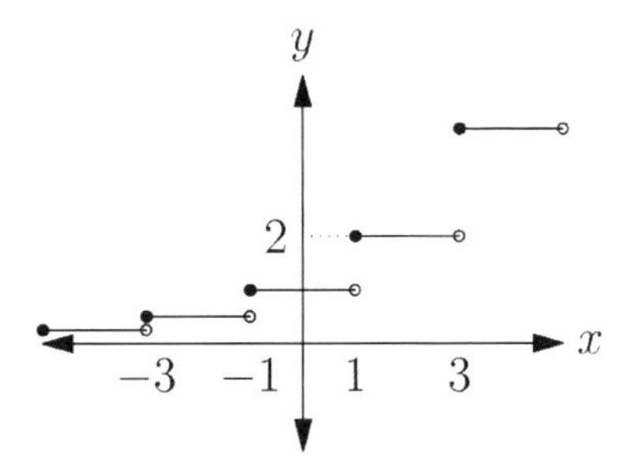


As the line segment of length 2 is pushed to the right by 2, then the height becomes twice the previous height, as one can check from the diagram above. Hence, the answer is **(A)**.

**5.**

All the answer choices are located in the first quadrant except $(1, -\frac{7\pi}{3})$ which stays in the fourth quadrant. Hence, the answer must be **(E)**.

**6.**

Let $x = 8 - t$ and $y = 10 + 2t$. Then, $2x = 16 - 2t$ and $y = 10 + 2t$. Add the two equations together to get $2x + y = 26$. Hence, $y = -2x + 26$. The slope of the line must be $-2$. The answer is **(B)**.

**7.**

Since $1 + \tan^2(t) = \sec^2(t)$ for any $t$,

$$\begin{aligned} x^2 - y^2 &= \sec^2(t) - \tan^2(t) \\ &= 1 \end{aligned}$$

This is an equation of a hyperbola, the difference of squares equaling 1. Hence, the answer is **(C)**.

**8.**

In order to find the $x$-intercept, we must set $y = 0$. Notice that $y = 0$ at $t = 4$. Therefore, $x(4) = 6 - 2(4) = 6 - 8 = -2$. The answer must be **(C)**.

**9.**

Let's use the distance formula between the two points.

$$\begin{aligned} \sqrt{(4t - 3 - 0)^2 + (0 - 2 + 4t)^2} &= \sqrt{16t^2 - 24t + 9 + 16t^2 - 16t + 4} \\ &= \sqrt{32t^2 - 40t + 13} \\ &= \sqrt{32\left(t^2 - \frac{40}{32}t\right) + 13} \\ &= \sqrt{32\left(t^2 - \frac{5}{4}t + \frac{25}{64} - \frac{25}{64}\right) + 13} \\ &= \sqrt{32\left(t^2 - \frac{5}{4}t + \frac{25}{64}\right) - \frac{25}{2} + 13} \\ &= \sqrt{32\left(t - \frac{5}{8}\right)^2 + \frac{1}{2}} \end{aligned}$$

The minimum distance of $\frac{1}{2}$ is reached by the two points at $t = \frac{5}{8} = 0.625$, so the answer is **(B)**.

**10.**

We will use the following conversion equations between polar forms and rectangular forms of coordinates.

- $x = r\cos(\theta)$
- $y = r\sin(\theta)$
- $r^2 = x^2 + y^2$

$$\begin{aligned} r &= \frac{1}{1+\cos(\theta)} \\ r(1+\cos(\theta)) &= 1 \\ r + r\cos(\theta) &= 1 \\ r &= 1 - r\cos(\theta) \\ r^2 &= (1 - r\cos(\theta))^2 \\ x^2 + y^2 &= (1-x)^2 \\ x^2 + y^2 &= 1 - 2x + x^2 \\ y^2 &= 1 - 2x \end{aligned}$$

The equation above represents a parabola because there is only one square term. Hence, the answer is **(C)**.

# Topic 11

# Counting and Probability

- ✓ Principles of addition or multiplication
- ✓ Inclusion and exclusion principle
- ✓ Geometric counting
- ✓ Probability

## 11.1 Principles of addition and multiplication

We either **add** or **multiply** when we count.

- If "$A$, then $B$", we (**multiply** / add).
- If "either $A$ or $B$, we (multiply / **add**).

---

**MATHEMATICS LEVEL 2 Test - *Continued***

| $T$ | $E$ | $A$ | $C$ | $U$ | $P$ |
|---|---|---|---|---|---|

USE THIS SPACE FOR SCRATCH WORK.

1. Given six cards $T$,$E$,$A$,$C$,$U$,$P$, shown in the diagram above. If six cards are listed in an array, consonants must be placed in the odd position, whereas vowels must be placed in the even position. Which of the following is the total number of arrangements?

(A) 30
(B) 32
(C) 34
(D) 36
(E) 38

2. How many six-digit even numbers can be formed from the numbers $0, 1, 2, 3, 4, 5$ if no digit is repeated?

(A) 96
(B) 216
(C) 312
(D) 360
(E) 720

- We ***multiply*** when two events are **successive**.
- We also ***multiply*** if calculations are **repetitive**.
- We ***add*** when two events **do not happen** at the same time.
- We ***add*** if calculations are **different**.

---

**MATHEMATICS LEVEL 2 Test - *Continued***

USE THIS SPACE FOR SCRATCH WORK.

3. Which of the following is the total number of positive integer triplets $(x, y, z)$ if $3x + 2y + z = 12$?

(A) 1
(B) 2
(C) 4
(D) 7
(E) 11

4. If a die is rolled three times in a row, let $a$, $b$, and $c$ be the numbers appearing on a die each time we throw. Which of the following counts the number of cases when $\overline{abc}$ is a multiple of 9?

(A) 25
(B) 26
(C) 27
(D) 28
(E) 29

## 11.2 Inclusion and exclusion principle

- $n(A \cup B) = n(A) + n(B) - n(A \cap B)$
- $n(A \cap B) = n(B) \times n(A|B)$
- $n(B|A) = n(B)$ if $A$ and $B$ are **independent**. Normally, the subject test assumes two events are independent.
- We ***subtract*** if events are **overcounted**.

---

**MATHEMATICS LEVEL 2 Test - *Continued***

USE THIS SPACE FOR SCRATCH WORK.

5. Out of whole numbers from 1 to 30 inclusive, which of the following counts the total number of multiples of 2 or 3?

(A) 10
(B) 15
(C) 20
(D) 25
(E) 30

6. Given natural numbers ranging from 1 to 9999 inclusive, which of the following is the total number of 7 appearing in the list?

(A) 1000
(B) 2000
(C) 3000
(D) 4000
(E) 5000

**MATHEMATICS LEVEL 2 Test - *Continued***

USE THIS SPACE FOR SCRATCH WORK.

7. Suppose there are 7-digit numbers that consist of 2 and 3 only. Which of the following is the number of expressions that has one 2332? (Note: 2332332 has two 2332's in the expression.)

(A) 8
(B) 16
(C) 23
(D) 30
(E) 37

8. How many multiples of 4 or 6 are there for all natural numbers ranging from 1 to 100, inclusive?

(A) 16
(B) 25
(C) 32
(D) 41
(E) 49

## 11.3 Geometric counting

- Triangle is formed given three **noncollinear** points.
- Out of $n$ objects, choosing 2 objects is equal to $\binom{n}{2} = \frac{n(n-1)}{2}$.

**MATHEMATICS LEVEL 2 Test - *Continued***

USE THIS SPACE FOR SCRATCH WORK.

9.

How many non-congruent segments can be drawn from these five points?

(A) 4
(B) 5
(C) 10
(D) 15
(E) 20

10. There are $n$ points on a circle. The number of pentagons formed out of these $n$ points equals the number of hexagons formed out of the $n$ points on the circle. What is the value of $n$?

(A) 5
(B) 6
(C) 11
(D) 30
(E) 41

**MATHEMATICS LEVEL 2 Test - *Continued***

USE THIS SPACE FOR SCRATCH WORK.

11. If two lines $l$ and $m$ are parallel, how many triangles can be formed using the points in the following diagrams?

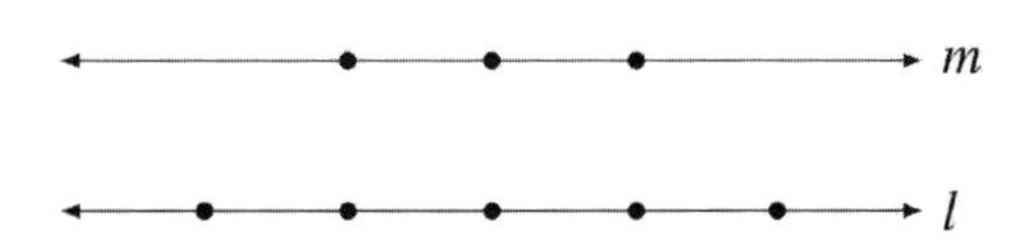


(A) 15
(B) 30
(C) 45
(D) 60
(E) 75

## 11.4 Probability

Probability is all about proportion. Probability of an event $A$ is the ratio of the number of ways $A$ can happen to the total number of outcomes.

Counting mechanism follows both principles of addition and multiplication. If events $A$ or $B$ cannot occur at the same time, we simply add probabilities. On the other hand, if events $A$ and $B$ can happen *simultaneously* or *successively*, then we multiply the probabilities.

- If you see the phrase **at least**, perform **complementary counting**.
- If you see the phrase similar to **given that**, then the **sample space** shrinks.[1]

**MATHEMATICS LEVEL 2 Test - *Continued***

USE THIS SPACE FOR SCRATCH WORK.

12. A utensil case contains three yellow pens, one blue pen, and two red pens. There are also two big erasers and one small one. If Bob randomly chooses one pen and one eraser, what is the probability of Bob's getting the blue pen with the small eraser?

(A) $\frac{1}{5}$ (B) $\frac{1}{15}$ (C) $\frac{3}{5}$ (D) $\frac{1}{3}$ (E) $\frac{1}{2}$

13. If Henry rolls a fair die and flips a coin, what is the probability of rolling a number less than 5 on the die and flipping heads on the coin?

(A) $\frac{1}{3}$ (B) $\frac{1}{2}$ (C) $\frac{1}{4}$ (D) $\frac{2}{3}$ (E) $\frac{2}{5}$

[1] Sample space is the set of all possible outcomes.

## MATHEMATICS LEVEL 2 Test - *Continued*

USE THIS SPACE FOR SCRATCH WORK.

14. If two fair dice are thrown, what is the probability that one die was four, given that the sum of values is seven?

(A) $\frac{1}{6}$ (B) $\frac{1}{5}$ (C) $\frac{1}{4}$ (D) $\frac{1}{3}$ (E) $\frac{1}{2}$

15. There are two alarm clocks. One of the alarm clock goes off at the right time with the probability of 0.95. The other alarm clock rings at the right time with the probability of 0.92. If the two events are independent, what is the probability that at least one alarm clock goes off?

(A) 0.996
(B) 0.945
(C) 0.915
(D) 0.865
(E) 0.755

## Answer Key to Practice Problems

1. (D)

2. (C)

3. (D)

4. (B)

5. (C)

6. (D)

7. (D)

8. (D)

9. (A)

10. (C)

11. (C)

12. (A)

13. (A)

14. (D)

15. (A)

## Detailed Solution for Practice Problems

### 1.

There are three vowels, $a, e, u$. Let's place these three letters in odd positions in 3! ways. The rest three letters are consonants, and the number of ways to place these consonants in even positions is 3!. We placed "vowels, then consonants," so we multiply $3! \times 3! = 36$. We multiply when events are chained up by discrete actions, then count the number of arrangement of consonants." On the other hand, we add if we do the casework. Thus, the answer is **(D)**.

### 2.

Let's case-enumerate. The principle of caseworks is simple : disjointly and completely. No matter what you do, when you casework, you have to find all possible cases that are disjoint, meaning that the events cannot overlap.

Case 1. The last digit is 0. Then, all the other letters can be arranged in $5! = 120$ ways.

Case 2. The last digit is 2. Then, 0 cannot be placed in the first digit, so there are $4 \times 4 \times 3 \times 2 \times 1 = 96$ ways.

Case 3. The last digit is 4. Just like in case 2, 0 cannot be placed in the first digit, so there are $4 \times 4! = 96$ ways.

In case enumeration, we add the number of counts in case 1, 2, and 3, so the answer is **(C)**.

### 3.

Let's do the casework for values of $x$. You could have case-enumerated for other variables, but I simply chose $x$ because there are fewer cases to cover.

Case 1. $x = 1$. Then, $2y + z = 9$, so $(y, z) = (1,7), (2,5), (3,3), (4,1)$.

Case 2. $x = 2$. Then, $2y + z = 6$, so $(y, z) = (1,4), (2,2)$.

Case 3. $x = 3$. Then, $2y + z = 3$, so $(y, z) = (1,1)$.

Adding all possible triples, we get the answer **(D)**.

**4.**

We do the casework again. Remember that the principle of caseworks is simple : disjointly and completely.

Case 1. $a+b+c=9$. Let's try to compute the number of triples satisfying $a \leq b \leq c$, for disjointness and completeness, and compute other variations.

- $(a,b,c)=(1,2,6)$. There are 3! possible arrangements, i.e., $(a,b,c)=(1,2,6),(1,6,2),(2,1,6),\cdots,(6,2,1)$.
- $(a,b,c)=(1,3,5)$. There are 3! possible arrangements, i.e., $(a,b,c)=(1,3,5),(1,5,3),\cdots,(5,3,1)$.
- $(a,b,c)=(1,4,4)$. There are 3 possible arrangements, i.e. $(a,b,c)=(1,4,4),(4,1,4),(4,4,1)$.
- $(a,b,c)=(2,2,5)$. There are 3 possible arrangements, i.e., $(a,b,c)=(2,2,5),(2,5,2),(5,2,2)$.
- $(a,b,c)=(2,3,4)$. There are 3! possible arrangements, i.e., $(a,b,c)=(2,3,4),(2,4,3),\cdots,(4,3,2)$.
- $(a,b,c)=(3,3,3)$. There is 1 possible arrangement.

Case 2. $a+b+c=18$. There is only one possible arrangement, i.e., $(a,b,c)=(6,6,6)$.

To sum up, there are 25 possible cases when $\overline{abc}$ is a multiple of 9. Hence, the answer must be **(B)**.

**5.**

This question uses the principle of inclusion and exclusion. First, let's count the number of multiples of 2 and 3.

- The number of multiples of 2 can be computed as $\left\lfloor \frac{30}{2} \right\rfloor = 15$.
- The number of multiples of 3 can be computed as $\left\lfloor \frac{30}{3} \right\rfloor = 10$.

However, the answer is less than 25 because there are *overcounts*. We counted the multiples of 6 twice, so we need to eliminate the overcounts. The number of multiples of 2 and 3 can be computed as $\left\lfloor \frac{30}{6} \right\rfloor = 5$. Therefore, the total number of multiples of 2 or 3 is $15+10-5=20$. The answer is **(C)**.

**6.**

We will do the casework, again.

Case 1. 7□□□ : There are $10^3(=1,000)$ possible cases with the thousands digit as 7.

Case 2. □7□□ : There are $10^3(=1,000)$ possible cases with the hundreds digit as 7.

Case 3. □□7□ : There are $10^3(=1,000)$ possible cases with the tens digit as 7.

Case 4. □□□7 : There are $10^3(=1,000)$ possible cases with the units digit as 7.

Adding all up, we get 4,000 number of 7's appearing in the list. You might wonder why I did not exclude any of the overcounts. Well, in this case, I have to include all overcounts. For instance, have a look at 7777. How many 7's are used in the number? Your answer should be 4 times. In case 1, there has been one count for 7777. In case 2, there is another count of 7777, and so on. In this question, we do not get rid of any overcounts. In fact, it is crucial not to exclude any of the overcounts. The answer must be **(D)**.

**7.**

Another practice for casework. In the question, there is one condition that 1221221 has two 1221's in the expression. We should delete this case from our counts as we count with the casework.

Case 1. 1221□□□ : $2^3-1(=7)$ possibilities. Think about why I deleted one from eight possibilities. That one should be 1221221.

Case 2. □1221□□ : $2^3(=8)$ possibilities. First, of all, we can place 2 values into the first square. Likewise, there are 2 possible choices for the second square. Lastly, there are 2 choices for the last square. We multiply, not add, in this case, so the computation turns into 8, not 6.

Case 3. □□1221□ : $2^3(=8)$ possibilities. Every counting mechanism in case 3 is exactly same as the one in case 2.

Case 4. □□□1221 : $2^3-1(=7)$ possibilities. Just as in case 1, I should delete one case of 1221221.

In total, there are $2\times7+2\times8(=30)$ expressions that has one 1221. The answer is **(D)**.

**8.**

We replicate the solution for question 5.

- The number of multiples of 4 is $\left\lfloor \frac{100}{4} \right\rfloor = 25$.
- The number of multiples of 6 is $\left\lfloor \frac{100}{6} \right\rfloor = 16$.
- The number of multiples of 12, the least common multiple of 4 and 6, is $\left\lfloor \frac{100}{12} \right\rfloor = 8$.

Therefore, $25 + 16 - 8 = 41 - 8 = 33$ multiples of 4 or 6 are between 1 and 100, inclusive. Hence, the answer is **(D)**.

**9.**

Line segments are congruent if the lengths are equal. Hence, we are looking for different segments. There are only four segments with different side lengths, so the answer is **(A)**.

**10.**

This is an application of the property of combination. Suppose there are 10 students in a class. I, as a teacher, wish to select two students to do details after the class. The number of selection is exactly equal to the number of selecting eight students to remain in the class and not do details. In mathematics, we write

$$\binom{n}{r} = \binom{n}{n-r}$$

In this question, we can apply it as $\binom{n}{5} = \binom{n}{6}$, where $n = 5 + 6 = 11$. Hence, the answer is **(C)**.

**11.**

If we use the casework, the counts are

$$\begin{aligned} \binom{3}{1} \times \binom{5}{2} + \binom{3}{2} \times \binom{5}{1} &= 30 + 15 \\ &= 45 \end{aligned}$$

Hence, the answer is **(C)**.

**12.**

- The probability of Bob choosing the blue pen : $\frac{3}{5}$
- The probability of Bob choosing the smaller eraser : $\frac{1}{2}$

We must decide whether we add or multiply the probabilities. Think about it this way. After computing the probability of Bob's choosing the blue pen, am I done with the whole computation? No. I need to consider the probability of Bob's choosing an eraser. In such case, we consider that it should be an application of multiplication principle, i.e., $\frac{3}{5} \times \frac{1}{3} = \frac{1}{5}$. In other words, if actions are incomplete, you keep multiplying the probabilities(or the number of counts). Hence the answer is **(A)**.

**13.**

The probability for Henry to roll a number less than 5 on the die is $\frac{4}{6}$. The probability for him to flip heads on the coin is $\frac{1}{2}$. We multiply the two probabilities, $\frac{4}{6} \times \frac{1}{2} = \frac{1}{3}$, so the answer is **(A)**.

**14.**

Let's write down all sample space $\{(1,6),(2,5),(3,4),(4,3),(5,2),(6,1)\}$. Out of these six possibilities, we need to count the proportion that one of the face values is 4. Hence, the answer must be $\frac{2}{6} = \frac{1}{3}$, which is **(D)**.

**15.**

If you see a phrase *at least*, then think about complementary counting, which eliminates the probabilities of unlikely events from 1. In our case,

$$\begin{aligned} 1-(1-0.95)\times(1-0.92) &= 1-(0.05)(0.08) \\ &= 0.996 \end{aligned}$$

In fact, we can do the casework.

- Alarm 1 goes off but alarm 2 does not : $0.95 \times 0.08$
- Alarm 2 goes off but alarm 1 does not : $0.92 \times 0.05$
- Alarm 1 and alarm 2 go off : $0.92 \times 0.95$

Adding all of them, we get 0.996, as well. The answer is **(A)**.

## Independent or Mutually Exclusive

$A$ and $B$ are independent if $p(A \cap B) = p(A) \times p(B)$.

Normally, $p(A \cap B) = p(A) \times p(B|A) = P(B) \times p(A|B)$. If two events are independent, then $p(A|B) = p(A)$ or $p(B|A) = p(B)$. In other words, regardless of whether one event happens, the probability of other event happening is unaffected.

$A$ and $B$ are mutually exclusive if $p(A \cap B) = 0$.

If $A$ and $B$ are mutually exclusive, then $p(A \cap B) = 0$. There is no common event between $A$ and $B$. The main difference between independent events and mutually exclusive events is that two independent events DO have something in common, whereas two mutually exclusive events DO NOT have anything in common.

For instance, if $p(A) = \frac{1}{3}$, $p(B) = \frac{1}{4}$, then

- Two events are independent if $p(A \cap B) = p(A) \times p(B) = \frac{1}{12}$.
- Two events are mutually exclusive if $p(A \cup B) = p(A) + p(B) = \frac{1}{3} + \frac{1}{4} = \frac{7}{12}$.

# Topic 12

# Geometry

- ✓ Circle equations
- ✓ Parabola, hyperbola and ellipse
- ✓ Polygons other than triangles
- ✓ Space geometry

## 12.1 Circle equations

Circle is the set of points **equidistant** from a fixed point $(h,k)$ called the center. The circle equation with center $(h,k)$ and radius $r$ is given by

$$(x-h)^2+(y-k)^2=r^2$$

If the point is inside the circle, we say the point is interior of it. On the other hand, if the point is outside the circle, then the point is outside of it.

- Area $=\pi r^2$
- Circumference $=2\pi r$

---

**MATHEMATICS LEVEL 2 Test - *Continued***

USE THIS SPACE FOR SCRATCH WORK.

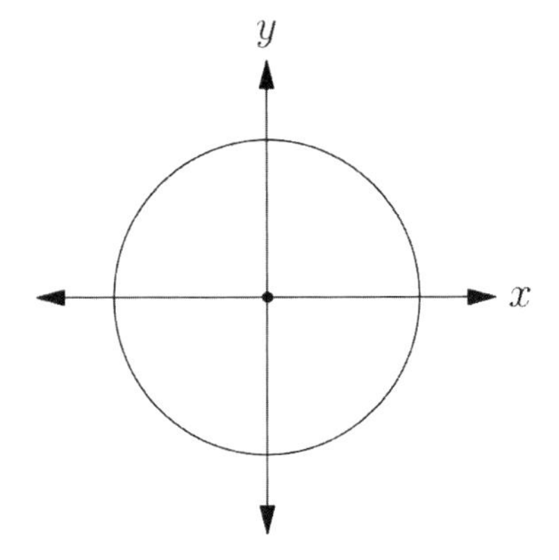


1. Which of the following is an interior point of a circle centered at the origin with the radius of 5?

(A) $(4,4)$
(B) $(5,3)$
(C) $(3,4)$
(D) $(2,3)$
(E) $(-3,-4)$

2. The radius of the circle $x^2+4x+y^2-10y-20=0$ is equal to

(A) 4 (B) 5 (C) 7 (D) 9 (E) 20

**MATHEMATICS LEVEL 2 Test - *Continued***

USE THIS SPACE FOR SCRATCH WORK.

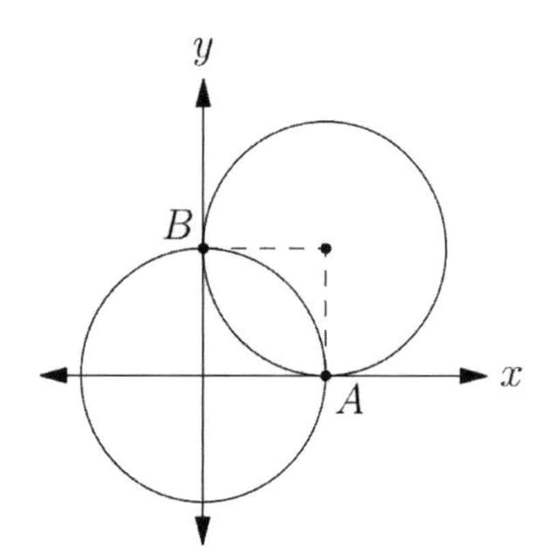


3. If $A$ and $B$ are the points of intersection of two circles whose equations are $x^2+y^2=9$ and $(x-3)^2+(y-3)^2=9$, what is the slope of the line perpendicular to the line that passes through $A$ and $B$?

(A) 2
(B) 1
(C) 0
(D) $-1$
(E) $-2$

4. If $x^2+ax+y^2+by+c=0$ is satisfied by $(3,1)$ and $(7,1)$, which of the following must be the value of $a$?

(A) 5
(B) 10
(C) 1
(D) $-10$
(E) $-5$

## 12.2 Parabola, hyperbola and ellipse

First, ellipse is the set of points whose sum of distances from the two points (known as foci) is fixed.

$$\frac{(x-h)^2}{a^2}+\frac{(y-k)^2}{b^2}=1$$

- $(h,k)$ is the center.
- If $|a|>|b|$, then $2|a|$ is the length of major axis.
- If $|a|>|b|$, then $2|b|$ is the length of minor axis.

Second, hyperbola is the set of points whose difference of distances from the two points (known as foci) is fixed.

$$\frac{(x-h)^2}{a^2}-\frac{(y-k)^2}{b^2}=1 \qquad \frac{(y-k)^2}{b^2}-\frac{(x-h)^2}{a^2}=1$$

- $(h,k)$ is the center.
- The point of intersections between the asymptotes $y=\pm\frac{b}{a}(x-h)+k$ is the center $(h,k)$.

Lastly, parabola is the set of points whose distance to a line equals that to a point.

$$4p(y-k)=(x-h)^2 \qquad 4p(x-h)=(y-k)^2$$

- $|p|$ is the distance between the vertex and the focus.
- If $y=x^2$, then parabola goes upward.
- If $x=y^2$, then parabola goes rightward.

---

**MATHEMATICS LEVEL 2 Test - *Continued***

5. What is the length of major axis of $\frac{x^2}{9}+\frac{y^2}{4}=1$?

USE THIS SPACE FOR SCRATCH WORK.

(A) 2 (B) 3 (C) 5 (D) 6 (E) 9

**MATHEMATICS LEVEL 2 Test - *Continued***

USE THIS SPACE FOR SCRATCH WORK.

6. If a hyperbola has two asymptotes at $y = 3x - 4$ and $y = 5x - 2$, what is the $y$-coordinate of its center?

(A) $-1$
(B) 7
(C) $-7$
(D) 1
(E) 0

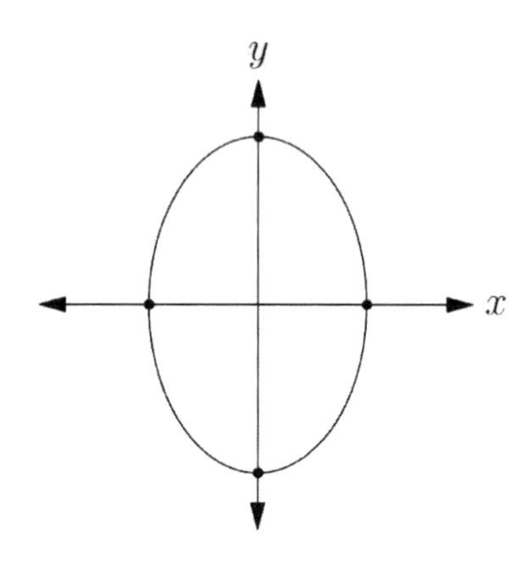


7. What is the length of the major axis of the ellipse $\frac{x^2}{10} + \frac{y^2}{20} = 1$?

(A) 3.2
(B) 4.5
(C) 8.9
(D) 10.0
(E) 20.0

On the other hand, the test also asks students about the definition of angle bisector and perpendicular bisector.

- Angle bisector is the set of points equidistant to two distinct intersecting lines.
- Perpendicular bisector is the set of points equidistant to two distinct points.

---

**MATHEMATICS LEVEL 2 Test - *Continued***

USE THIS SPACE FOR SCRATCH WORK.

8. Given two distinct non-parallel lines, which of the following is the set of points equidistant from the two lines?

(A) Two perpendicular lines.
(B) Two parallel lines.
(C) Two points.
(D) One line.
(E) One point.

9. Which of the following equations is the set of points equidistant from $(1,2,3)$ and $(3,2,3)$?

(A) $x = 1$
(B) $y = 2$
(C) $z = 3$
(D) $x = 2$
(E) $x = -2$

## 12.3 Polygons other than triangles

Other than triangles, basic knowledge of quadrilaterals is tested in the test along with some special types of questions.

- Parallelogram : Diagonals bisect each other.
- Rectangle : Congruent diagonals bisect each other.
- Rhombus : Diagonals are perpendicularly bisected.
- Square : Congruent diagonals are perpendicularly bisected.

Out of many types of questions, we can have a look at two special types of application of polygons.

- Similarity : if the scale factor is $a : b$, then the area ratio is $a^2 : b^2$ for similar figures.
- Congruent polygons : Cutting a figure into two congruent parts is to use the center point of the polygon.

---

**MATHEMATICS LEVEL 2 Test - *Continued***

USE THIS SPACE FOR SCRATCH WORK.

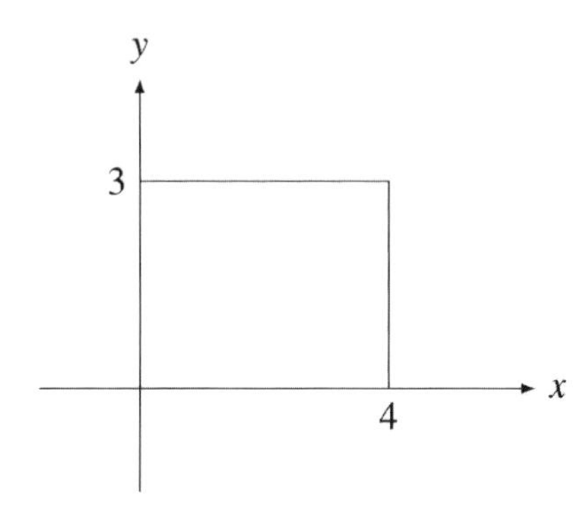


10. Which of the following equations of a line cut the rectangle above into two congruent parts?

(A) $y = \frac{1}{2}x + \frac{1}{2}$

(B) $y = \frac{1}{2}x - \frac{1}{2}$

(C) $y = -\frac{3}{4}x + 1$

(D) $y = -\frac{4}{3}x - 1$

(E) $y = \frac{2}{3}x + \frac{1}{2}$

**MATHEMATICS LEVEL 2 Test - *Continued***

11. Given a triangle $ABC$ with $A(2,1)$, $B(5,5)$ and $C(5,1)$, if all edges of the triangle are stretched by a factor of 2, then how many times larger is the area of a new triangle than that of the original?

USE THIS SPACE FOR SCRATCH WORK.

(A) 2 (B) 4 (C) 6 (D) 8 (E) 16

12. If a diagonal of a rhombus is on the $x$-axis with the endpoints of $(-4.56, 0)$ and $(4.56, 0)$, what is the sum of all coordinates of its vertices?

(A) 4.56
(B) $-9.12$
(C) 0
(D) 9.12
(E) $-4.56$

## 12.4 Space Geometry

Solids are easy to identify, but the terminologies are difficult to remember. The following list labels the name for each solid.

- Tetrahedron : solids with four faces
- Cube : solids with six faces
- Octahedron : solids with eight faces
- Dodecahedron : solids with twelve faces
- Icosahedron : solids of twenty faces

Other than the solids listed above, there are commonly-used solids for students to figure out surface area and volume.

- Cuboid : Rectangular cube
- Cone : Circular pyramid (rotation of a right triangle at its side as the axis of rotation)
- Cylinder : Circular prism (rotation of a rectangle at its side as the axis of rotation)
- Sphere : a misnomer for 3D ball. Sphere is actually a surface of a ball. In highschool, sphere refers to a ball.
- Hemisphere : a half-sphere

For cube, cuboid, cone, and cylinder, there are special questions related to the longest length. It is a longest diagonal question. Not only is it a direct application of Pythagorean Theorem but also it is a corresponding application of distance formula in 3-dimension.

- Cube : the length of the longest diagonal is $x\sqrt{3}$ where $x$ is the length of an edge.
- Cuboid : The length of the longest diagonal is $\sqrt{x^2+y^2+z^2}$ where $x$, $y$, and $z$ are side lengths.
- Cone : The length of slanted height is $\sqrt{r^2+h^2}$ where $r$ is the base radius and $h$ is the height.
- Cylinder : The length of the longest possible diagonal is $\sqrt{(2r)^2+h^2}$ where $r$ is the base radius and $h$ is the height.

For 3-dimensional figures, the surface area and volume equations are easy to remember, all of which are listed below.

- Prism : $V = Bh$ and $SA = 2B + 4L$ where $B$ is the base area and $L$ is the lateral area.
- Pyramid : $V = \frac{1}{3}Bh$ and $SA = 3\triangle B$ where $\triangle$ is the lateral area and $B$ is the base area.
- Cylinder : $V = \pi r^2 h$ and $SA = 2\pi rh + 2\pi r^2$ where $r$ is the base radius and $h$ is the height.
- Cone : $V = \frac{1}{3}\pi r^2 h$ and $SA = \pi rl + \pi r^2$ where $l$ is the slant height and $r$ is the radius.
- Cube : $V = x^2$ and $SA = 6x^2$ where $x$ is the length of an edge.
- Cuboid : $V = xyz$ and $SA = 2(xy + yz + zx)$ where $x$, $y$, and $z$ are side lengths.
- Sphere : $V = \frac{4}{3}\pi r^3$ and $SA = 4\pi r^2$ where $r$ is the radius.

Along with these volume and surface area questions, the test also includes the locus questions. A locus is a set of points satisfying a certain condition.

- The locus of points at a fixed distance $d$ from a point $P$ is a circle of radius $d$ with the center at $P$. In other words, a locus of points whose distance from a point is fixed is a circle.
- The locus of points at a fixed distance $d$ from a line $L$ is two parallel lines $d$ distance from $L$ to either side. In other words, a fixed distance from a line results in two parallel lines.
- The locus of points equidistant from two points $A$ and $B$ is the perpendicular bisector of the line segment $\overline{AB}$. In other words, the equidistance condition from two distinct points results in a perpendicular bisector.
- The locus of points equidistant from two parallel lines, $L_1$ and $L_2$, is a line parallel to both $L_1$ and $L_2$, in the middle of the two. In other words, the equidistance condition from two parallel lines results in a parallel line in the middle.
- The locus of points equidistant from two intersecting lines $L_1$ and $L_2$ is a pair of bisectors bisecting the angles formed by $L_1$ and $L_2$. In other words, the equidistance condition from two intersecting lines results in two perpendicular lines.

**MATHEMATICS LEVEL 2 Test - *Continued***

USE THIS SPACE FOR SCRATCH WORK.

13. If $X$ and $Y$ are different points in a plane, the set of all points in this plane closer to $X$ then to $Y$ is

(A) a region of the plane on the side of a line that contains $X$.
(B) the interior of a circle
(C) a U-shaped region in the plane
(D) the region of the plane bounded by a parabola.
(E) the interior of a square

14. A diagonal of the base of the prism has the length of $\sqrt{47}$, and the prism has the height of 3. Which of the following is the length of the diagonal of the prism?

(A) $\sqrt{42}$
(B) $\sqrt{47}$
(C) $\sqrt{56}$
(D) $\sqrt{81}$
(E) $\sqrt{141}$

**MATHEMATICS LEVEL 2 Test - *Continued***

USE THIS SPACE FOR SCRATCH WORK.

15. The intersection of a cube with a plane could be which of the following?

I. a square

II. a parallelogram

III. a triangle

(A) None of the above
(B) I and II
(C) I and III
(D) II and III
(E) I, II, and III

16. The vertices of a rhombus $P(a, b)$ is moved to $P(2a, 2b)$. What is the ratio between the area of new rhombus to that of the original one?

(A) 2 (B) 4 (C) $\sqrt{2}$ (D) $\frac{1}{2}$ (E) $\frac{1}{4}$

**MATHEMATICS LEVEL 2 Test - *Continued***

USE THIS SPACE FOR SCRATCH WORK.

17. The length, width, and height of a rectangular solid are 8, 4, and 1, respectively. What is the length of the longest line segment whose end points are two vertices of this solid?

(A) 8 (B) 9 (C) 10 (D) 11 (E) 12

18. If the height of a right circular cone is decreased by 8 percent, by what percent must the radius of the base be decreased so that the volume of the cone is decreased by 15 percent?

(A) 4 percent
(B) 5 percent
(C) 6 percent
(D) 7 percent
(E) 8 percent

19. The cone has a volume of 250 cubic units. If the height of the cone is 8 units, what is the radius of the cup?

(A) 4.25
(B) 4.65
(C) 4.95
(D) 5.26
(E) 5.46

## MATHEMATICS LEVEL 2 Test - *Continued*

20. There is a line in space perpendicular to a plane. The locus of all points that the distance from the plane is 4 and the distance from the line is 3 is two circles. How far is any point on the two circles from the point of intersection between a line and a plane?

(A) 3
(B) 4
(C) 5
(D) 6
(E) 7

USE THIS SPACE FOR SCRATCH WORK.

## Answer Key to Practice Problems

1. (D)

2. (C)

3. (B)

4. (D)

5. (D)

6. (C)

7. (C)

8. (A)

9. (D)

10. (A)

11. (B)

12. (C)

13. (A)

14. (C)

15. (E)

16. (B)

17. (B)

18. (A)

19. (E)

20. (C)

## Detailed Solution for Practice Problems

**1.**

The circle centered at the origin with the radius of 5 has the equation $x^2+y^2=25$. If the point is inside the circle, then $x^2+y^2<25$. Out of all answer choices, $(2,3)$ is the only point that satisfies this inequality, so the answer must be **(D)**.

**2.**

$$\begin{aligned} x^2+4x+y^2-10y-20&=0 \\ (x^2+4x+4)+(y^2-10y+25)-20&=4+25 \\ (x-2)^2+(y-5)^2&=49 \end{aligned}$$

Since $49=7^2$, the radius must be 7. Hence, the answer is **(C)**.

**3.**

Substituting $x^2+y^2=9$ into $(x-3)^2+(y-3)^2=9$, we get

$$\begin{aligned} (x-3)^2+(y-3)^2&=x^2+y^2 \\ x^2-6x+9+y^2-6y+9&=x^2+y^2 \\ -6x-6y+18&=0 \\ 18&=6x+6y \\ 3&=x+y \\ y&=-x+3 \end{aligned}$$

The line $y=-x+3$ passes through $A$ and $B$, so the slope of the line perpendicular to it is 1. The answer is **(B)**.

**4.**

Instead of substituting, we would like to use the fact that the perpendicular bisector of a chord always passes through the center. The chord connecting $(3,1)$ and $(7,1)$ is perpendicularly bisected by $x=5$. Hence, the circle equation must be $(x-5)^2+\cdots$. Hence, $x^2-10x+25+\cdots$ shows that $a=-10$. Thus, the answer is **(D)**.

**5.**

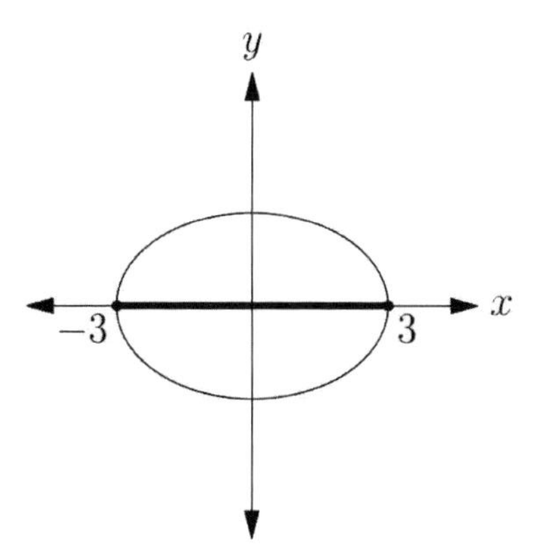


The length of the major axis is the length of the line segment connecting two vertices, not co-vertices. In our case, the length is 6, so the answer is **(D)**.

**6.**

The center of a hyperbola is located at the intersection point between the two asymptotes. Hence, $3x-4=5x-2$ implies that $x=-1$. Therefore, $y=3(-1)-4=-7$. The answer is **(C)**.

**7.**

Instead of graphing the ellipse, let's find the major axis length algebraically.

$$\frac{x^2}{(\sqrt{10})^2}+\frac{y^2}{(\sqrt{20})^2}=1$$

where $2\sqrt{20}\approx 8.9$ is the length of major axis. Therefore, the answer is **(C)**.

**8.**

The set of points equidistant from two distinct non-parallel lines is two angle bisectors that are perpendicular, so the answer is **(A)**.

**9.**

$$\sqrt{(x-1)^2+(y-2)^2+(z-3)^2}=\sqrt{(x-3)^2+(y-2)^2+(z-3)^2}$$
$$(x-1)^2=(x-3)^2$$
$$x^2-2x+1=x^2-6x+9$$
$$4x=8$$
$$x=2$$

Hence, the answer is **(D)**.

**10.**

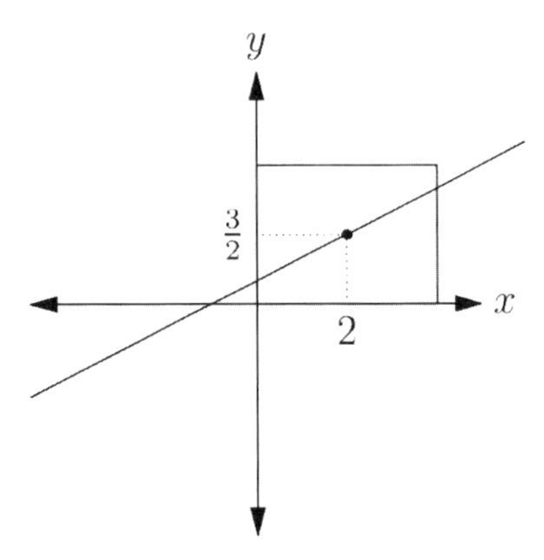


The graph of a line must pass through the midpoint of the diagonal of a rectangle $(2, \frac{3}{2})$, and the only line equation that passes through this point is **(A)**.

**11.**

Triangle $ABC$ and $A'B'C'$ must be similar with a scale factor of 2. Then, the area ratio between the two similar triangles must be $1:4$. Hence, the answer must be **(B)**.

**12.**

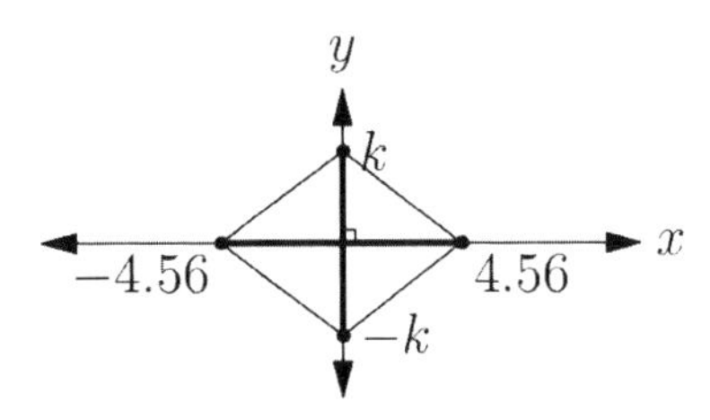


As one can see from the figure above, the rhombus has its diagonals perpendicularly bisected. Hence, the coordinates must be completely symmetric about the origin, which happens to be the center. Therefore, the sum of all coordinates should be 0, so the answer is **(C)**.

**13.**

This is an application of perpendicular bisector. The set of points equidistant from two distinct points is a perpendicular bisector. Hence, if we have to find the set of points whose distance from $X$ is smaller than that from $Y$, then we color the region of the plane that contains $X$, outlined by the perpendicular bisector. Hence, the answer is **(A)**.

**14.**

This is an application(or extension) of Pythagorean theorem. Specifically, if we let the length of the diagonal of the prism as $x$, then

$$\begin{aligned} x &= \sqrt{(\sqrt{47})^2 + 3^2} \\ &= \sqrt{56} \end{aligned}$$

Hence, the answer is **(C)**.

**15.**

Cutting a cube into half, we get a square. Since a square is a parallelogram, II is automatically true. Also, cutting the corner of the cube results in a triangle, so all three are true. Hence, the answer is **(E)**.

**16.**

The length ratio between two similar rhombuses is 2, so the area ratio must be the square of 2, i.e., 4. Hence, the answer is **(B)**.

**17.**

Just like question 14, we use Pythagorean theorem.

$$\begin{aligned} \sqrt{8^2 + 4^2 + 1^2} &= \sqrt{64 + 16 + 1} \\ &= \sqrt{81} \\ &= 9 \end{aligned}$$

The answer is **(B)**.

**18.**

$$\begin{aligned} \frac{1}{3} \times \pi(0.92h) \times (\square r)^2 &= 0.85V \\ 0.92 \times (\square)^2 &= 0.85 \\ (\square)^2 &= \frac{0.85}{0.92} \\ (\square)^2 &= 0.9239 \\ \square &= 0.961 \end{aligned}$$

Hence, the radius must be decreased roughly by 4 percent. Therefore, the answer is **(A)**.

**19.**

$$\frac{1}{3} \times \pi \times r^2 \times h = 250$$
$$\frac{1}{3} \times \pi \times r^2 \times 8 = 250$$
$$r^2 = \frac{250 \cdot 3}{8\pi}$$
$$r = \sqrt{\frac{750}{8\pi}}$$
$$r \approx 5.46$$

Therefore, the answer is **(E)**.

**20.**

Imagine we look at the plane horizontally. Then, the view can be illustrated as in the following diagram.

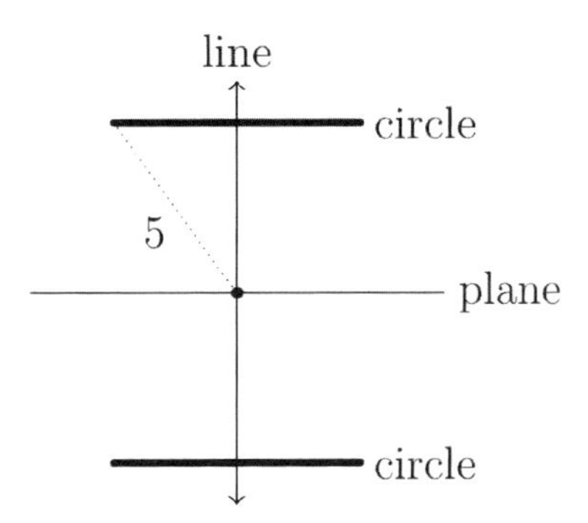


The distance between the point of intersection between the plane and the line and any point on the circle is 5, as shown in the figure above. Since two legs of a right triangle have the length of 3 and 4, respectively, the diagonal, shown as the dotted line segment, has the length of 5. Thus, the answer is **(C)**.

# Topic 13

# Numbers and Operations 1

- ✓ Binary Operation
- ✓ Elementary Number Theory
- ✓ Integer Equations

## 13.1 Binary Operation

Binary operation is an abstract notion used in Algebra. In spite of being abstract, it is easy to understand the concept. Simply, it is an operation of combining two elements by a binary operator. For example, + is a classic example of binary operation. Usually, "number + number" results in a number. Usually it is written that it has a closure under addition.

Now, in the subject test, this is more of an abstract notion. There is no fixed rule to remember for binary operation. Nevertheless, follow the given rule for any two values of number.

---

**MATHEMATICS LEVEL 2 Test -** ***Continued***

USE THIS SPACE FOR SCRATCH WORK.

1. If $x \bigstar y = xy - x + y^2$ for any $x$, $y$, what is the value of $3 \bigstar 5$?

(A) 40
(B) 25
(C) 37
(D) 43
(E) 49

2. If $a \triangle b = ab - a - b + 1$, which of the following expression results in 0?

(A) $(2,2)$
(B) $(0,2)$
(C) $(2,0)$
(D) $(1,1)$
(E) $(0,0)$

## 13.2 Elementary Number Theory

When we deal with integers, we solve it using the integer properties. Let's have a look at parity. If we denote even number as $E$ and odd as $O$, arithmetic rules for even and odd numbers can be given as

- $E \pm E = O \pm O = E$
- $O \pm E = O$
- $E \times E = E \times O = E$
- $O \times O = O$

**MATHEMATICS LEVEL 2 Test -** ***Continued***

USE THIS SPACE FOR SCRATCH WORK.

3. Given any consecutive array of three positive integers, what must be true?

I. the product is always even.

II. the sum is always even.

III. the sum is always odd.

(A) I only
(B) II only
(C) III only
(D) I and II only
(E) I and III only

4. The positive integers $M$, $N$, $N-M$, and $N+M$ are all distinct prime numbers. If $M$ is the smallest of all, what must be true about $M$?

(A) $M$ is even.
(B) $M$ is divisible by 3.
(C) $M$ is divisible by 5.
(D) $M$ is divisible by 7.
(E) $M$ is divisible by 11.

Now, let's have a look at multiples and factors.

- If $x$ is a multiple of $y$, then $x = ky$ for some $k$ integer.
- If $x$ is a factor of $y$, then $y = kx$ for some $k$ integer.

How about the number of positive divisors? In order to find the number of factors, prime factorize the given number. Add one to the powers and find the product of powers.

$$n = p^{\alpha} q^{\beta} \text{ where } p \text{ and } q \text{ are primes.}$$

Then, the number of positive divisors is $(\alpha + 1)(\beta + 1)$. For instance, given $36 = 2^2 3^2$, the number of positive factors of 36 is $(2+1)(2+1) = 9$. Here, prime number $p \geq 2$ is a positive integer whose positive factors are 1 and $p$. There are infinitely many prime numbers such that

$$p = \begin{cases} 2 \\ 3 \\ 6k \pm 1 \text{ for some } k \end{cases}$$

How about least common multiples and greatest common divisors? Least common multiple is the smallest common multiple of two or more integers. Greatest common divisor is the largest common factor of two or more integers. In order to find LCM or GCD, first find prime factorization of numbers, then choose either maximum or minimum exponent. For instance, given $m = a^p b^q c^r$ and $n = a^x b^y c^z$, then

- $\text{lcm}(m, n) = a^{\max(p,x)} b^{\max(q,y)} c^{\max(r,z)}$
- $\text{gcd}(m, n) = a^{\min(p,x)} b^{\min(q,y)} c^{\min(r,z)}$

---

**MATHEMATICS LEVEL 2 Test - *Continued***

5. The number $x$ is a three-digit positive integer and is the product of the three distinct prime factors $a$, $b$ and $10a + b$, where $a$ and $b$ are single-digit prime numbers. What is the largest possible value of $x$?

USE THIS SPACE FOR SCRATCH WORK.

(A) 533 (B) 777 (C) 795 (D) 957 (E) 997

Lastly, the test covers divisibility tests. The following rules sum up the divisibility properties of the first few positive integers.

- 2 : the units digit is even.
- 3 : the sum of all digits is divisible by 3 or 9.
- 4 : the last two digits of the number should be divisible by 4.
- 5 : the units digit must be either 0 or 5.
- 6 : the number should be divisible by both 2 and 3.
- 8 : the last three digits of the number should be divisible by 8.
- 9 : the sum of all digits is divisible by 9.
- 10 : the units digit must end with 0.

---

**MATHEMATICS LEVEL 2 Test - *Continued***

USE THIS SPACE FOR SCRATCH WORK.

6. Two distinct positive integers $x$ and $y$ are factors of 36. If the product of $x$ and $y$ does not divide 36, what is the smallest possible value of $x+y$?

(A) 4 (B) 5 (C) 6 (D) 8 (E) 12

7. If $ABC,123$ is a six-digit number that is divisible by 9, what could be the value of $A+B+C$?

(A) 2 (B) 3 (C) 4 (D) 5 (E) 6

## 13.3 Integer Equations

This might be particularly difficult to most students because one equation with two variables results in multiple solutions. In this section, we will approach this type of integer equations with systematic counting methods.

**MATHEMATICS LEVEL 2 Test - *Continued***

USE THIS SPACE FOR SCRATCH WORK.

8. If $5x+3y=20$, where $x$ and $y$ are non-negative integers, how many possible values of $x$ are there?

(A) 0
(B) 1
(C) 2
(D) 3
(E) 4

9. If there are 600 students seated on an auditorium in chairs with $m$ number of rows, each of which is occupied by $n$ number of students, which of the following CANNOT be the value of $m$?

(A) 15
(B) 20
(C) 24
(D) 25
(E) 45

**MATHEMATICS LEVEL 2 Test -** ***Continued***

USE THIS SPACE FOR SCRATCH WORK.

10. Two integers are co-prime if the common divisor between the two numbers is 1. If $n = p^3q^2r^2$, where $p$, $q$ and $r$ are distinct primes, how many divisors of $n$ are co-prime to $r$?

(A) 6
(B) 8
(C) 10
(D) 12
(E) 14

11. The least common multiple of $x$, 10 and 14 is 70. What is the greatest possible value of $x$?

(A) 10
(B) 14
(C) 35
(D) 70
(E) 140

12. Bob and Charles each selected a natural number less than 300. Bob chooses a multiple of 18, and Charles chooses a multiple of 24. Out of their choices, how many numbers are in common?

(A) 1
(B) 2
(C) 3
(D) 4
(E) 5

**MATHEMATICS LEVEL 2 Test - *Continued***

USE THIS SPACE FOR SCRATCH WORK.

13. If $a$ and $b$ are distinct odd primes, the number of positive divisors for the integer $a(2a+b)-2a^2+ab$ is equal to

(A) 2
(B) 4
(C) 6
(D) 8
(E) 10

14. How many positive integers less than or equal to 50 have an even number of positive divisors?

(A) 34
(B) 37
(C) 40
(D) 43
(E) 47

15. How many of the divisors of 7! are larger than 6!?

(A) 2
(B) 3
(C) 4
(D) 5
(E) 6

## Answer Key to Practice Problems

1. (C)

2. (D)

3. (A)

4. (A)

5. (C)

6. (C)

7. (B)

8. (C)

9. (E)

10. (D)

11. (D)

12. (D)

13. (D)

14. (D)

15. (E)

## Detailed Solution for Practice Problems

**1.**

$$\begin{aligned} 3\bigstar 5 &= (3)(5) - 3 + 5^2 \\ &= 15 - 3 + 25 \\ &= 40 - 3 \\ &= 37 \end{aligned}$$

The answer must be **(C)**.

**2.**

Since $ab - a - b + 1 = (a-1)(b-1)$, at least one of the values of $a$ and $b$ must be 1 in order to satisfy $a\triangle b = 0$. The only answer choice that satisfies this condition is **(D)**.

**3.**

If we call a triple of consecutive positive integers $(x, y, z)$, then $(x, y, z) = (1, 2, 3), (2, 3, 4), (3, 4, 5), (4, 5, 6), \cdots$. The product should be always even because it contains at least one even integer. On the other hand, the sum could be even or odd. Hence, the answer must be **(A)**.

**4.**

According to the condition, $N + M$ should be odd. Then, one of the values must be even, i.e., 2, and $M$ should be 2, because $N - M > 0$. Therefore, $M$ must be even. The answer is **(A)**.

**5.**

- $533 = 13 \times 41$
- $777 = 3 \times 7 \times 37$
- $795 = 5 \times 3 \times 53$
- $957 = 3 \times 11 \times 29$
- $997 = 1 \times 997$

The answer must be **(C)**.

**6.**

Let's write down all possible divisors of 36.

$$1, 2, 3, 4, 6, 9, 12, 18, 36$$

The smallest possible value of $x$ and $y$ such that $xy$ does not divide 36 is $(x, y) = (2, 4)$ or $(4, 2)$. Hence, $x + y = 6$, so the answer is **(C)**.

**7.**

Since $ABC, 123$ is a six-digit number divisible by 9, $A + B + C + 1 + 2 + 3$ should be divisible by 9. If $A + B + C = 3$, then $3 + 1 + 2 + 3 = 9$ is divisible by 9. Hence, the answer is **(B)**.

**8.**

If $x = 0$, then $3y = 20$. Since $y$ should be an integer, this is not valid. If $x = 1$, then $3y = 15$, so $y = 5$. If $x = 2$, then $3y = 10$, and the value of $y$ is not integer. If $x = 3$, then $3y = 5$, and this is invalid. If $x = 4$, then $3y = 0$, so $y = 0$. Therefore, there are two values of $x$ that result in integer values of $y$. Hence, the answer is **(C)**.

**9.**

- $600 = 15 \times 40$.
- $600 = 20 \times 30$.
- $600 = 24 \times 25$.
- $600 = 25 \times 24$.
- $600 = 45 \times 13\frac{1}{3}$.

The answer is **(E)**.

**10.**

We are looking for divisors of $n$ that has no power of $r$. This means that we are looking for divisors of $p^3q^2$, which is equal to $(3 + 1)(2 + 1) = 12$. Hence, the answer is **(D)**.

**11.**

- $10 = 2 \times 5$
- $14 = 2 \times 7$
- $70 = 2 \times 5 \times 7$

Hence, the greatest possible value of $x$ must be 70. Otherwise, the least common multiple is greater than 70, which leads to contradiction. Therefore, the answer is **(D)**.

**12.**

The least common multiple of $18(= 2 \times 3^2)$ and $24(= 2^3 \times 3)$ is $72(= 2^3 \times 3^2)$. Out of 300 natural numbers, we are looking for the number of multiples of 72, which is

$$\left\lfloor \frac{300}{72} \right\rfloor = 4$$

So, the answer is **(D)**.

**13.**

Since $a(2a+b) - 2a^2 + ab = 2a^2 + ab - 2a^2 + ab = 2ab = 2^1 a^1 b^1$, the number of positive divisors is $2^3 = 8$. Hence, the answer is **(D)**.

**14.**

If a number is a perfect square, then it has an odd number of positive divisors. Hence, out of 50, we eliminate $1, 4, 9, 16, 25, 36, 49$, so there are 43 numbers that have an even number of positive divisors. The answer must be **(D)**.

**15.**

Let $d > 6!$ be a divisor of $7!$. Then, $\frac{1}{d} < \frac{1}{6!}$. Since $d$ is a divisor of $7!$, $\frac{7!}{d} < \frac{7!}{6!} = 7$. Hence, $\frac{7!}{d} = 1, 2, 3, \cdots, 6$, so $d = 7!, \frac{7!}{2}, \frac{7!}{3}, \cdots, \frac{7!}{6}$. The answer must be **(E)**.

# Topic 14

# Numbers and Operations 2

- ✓ Conditional Statements
- ✓ Ratio and Proportion
- ✓ Sequence and Series
- ✓ Matrices

## 14.1 Conditional Statements

What is a conditional statement? Conditional statement consists of two mathematical statements : hypothesis and conclusion. It is referred as "if-then" statement, as well.

$$\text{if } P, \text{ then } Q.$$

We consider the statement as false if $P$ is false but $Q$ is true. In other cases, we consider it true. Let's have a look at stereotypical examples that teach us truth values of the conditional statement.

- if $x = 1$, then $x^2 = 1$ : This is true because the hypothesis only contains one value $x = 1$, whereas the conclusion contains two values $x = \pm 1$. In this case, the statement is always true.
- If $x^2 = 1$, then $x = 1$ : This is false because the hypothesis contains two values $x = \pm 1$ but the conclusion contains one value $x = 1$. Here, $x = -1$ is a counterexample that satisfies the hypothesis except the conclusion.

There are three associated conditional statements with the original one.

- Converse : if $Q$, then $P$.
- Inverse : if not $P$, then not $Q$.
- Contrapositive : if not $Q$, then not $P$.

Contrapositive and the original one share truth values. Similarly, converse and inverse share truth values.

---

**MATHEMATICS LEVEL 2 Test - *Continued***

1. If $p$ is prime, then $2p+1$ is a prime. Which value of $p$ is a COUNTEREXAMPLE to the mathematical statement?

USE THIS SPACE FOR SCRATCH WORK.

(A) 2
(B) 3
(C) 5
(D) 7
(E) 11

**MATHEMATICS LEVEL 2 Test - *Continued***

2. Which of the following must be true?

USE THIS SPACE FOR SCRATCH WORK.

I. If $x < 1$, then $x < 0$.

II. If $x^2 \neq 1$, then $x \neq 1$.

III. If $xy = 0$, then $x = 0$.

(A) I only
(B) II only
(C) III only
(D) I and III
(E) I, II, and III

3. All primes are odd. Which of the following values of prime is a COUNTEREXAMPLE to the given statement?

(A) 1
(B) 2
(C) 3
(D) 5
(E) 7

## 14.2 Ratio and Proportion

Two types of variations show up in the test. The first type is a direct variation between two variables.

$$y \text{ varies directly as } x. \longleftrightarrow y = kx \text{ for some constant } k.$$

The second type is an inverse variation between two variables.

$$y \text{ varies inversely as } x. \longleftrightarrow y = \frac{k}{x} \text{ for some constant } k.$$

---

**MATHEMATICS LEVEL 2 Test - *Continued***

USE THIS SPACE FOR SCRATCH WORK.

4. If $y$ varies inversely as the square root of $x$ where $x = 4$ and $y = 6$, then what is the value of $x$ when $y = 3$?

(A) 1
(B) 4
(C) 9
(D) 16
(E) 25

5. If $y$ varies directly as the square of $x$, and $x = 3$ if $y = 27$, what could be the value of $x$ when $y = 12$?

(A) 1
(B) 2
(C) 4
(D) 8
(E) 12

## 14.3 Sequence and Series

In Algebra II, students learned about two sequences : arithmetic and geometric. Arithmetic sequence is a sequence of terms with common difference. On the other hand, geometric sequence is a sequence of terms with common ratio. Now, let's have a look at the closed formula, which tells us the exact value for any value of $n$, for arithmetic sequence.

$$a_n = a_1 + (n-1)d$$

- $a_1$ is the first term.
- $d$ is the common difference.

Similarly, we have a formula for geometric sequence, i.e.,

$$b_n = b_1(r)^{n-1}$$

- $b_1$ is the first term.
- $r$ is the common ratio.

---

**MATHEMATICS LEVEL 2 Test - *Continued***

USE THIS SPACE FOR SCRATCH WORK.

6. If the fourth term of an arithmetic sequence is 10 and the seventh term is 16, what is the value of the first term?

(A) 2 (B) 4 (C) 6 (D) 8 (E) 10

7. If the fifth term of a geometric sequence is 20 and the seventh term is 5, what could be the value of the common ratio?

(A) 1 (B) 2 (C) 4 (D) $\frac{1}{2}$ (E) $-1$

The sum of sequential terms is called series. The sum of arithmetic sequence is given by the two formula.

$$S_n = \frac{n(a_1 + a_n)}{2}$$

where $a_1$ is the first term, $a_n$ is the last term, and $n$ is the total number of terms.

$$S_n = \frac{n(2a_1 + (n-1)d)}{2}$$

where $a_1$ is the first term, $d$ is the common difference, and $n$ is the total number of terms. On the other hand, the sum of geometric sequence is given by

$$S_n = \frac{a_1(1-r^n)}{1-r}$$

Specific to geometric series, the formula for infinite geometric series is given by

$$S = \frac{a_1}{1-r}$$

---

**MATHEMATICS LEVEL 2 Test - *Continued***

USE THIS SPACE FOR SCRATCH WORK.

8. The infinite geometric series $1 + \frac{1}{2} + \frac{1}{4} + \cdots$ equals

(A) 2 (B) 3 (C) 4 (D) 5 (E) 6

9. The sum of the first hundred positive integers, i.e., $1 + 2 + \cdots + 100$, is equal to

(A) 4000
(B) 4050
(C) 5000
(D) 5050
(E) 6000

**MATHEMATICS LEVEL 2 Test - *Continued***

USE THIS SPACE FOR SCRATCH WORK.

10. What is the arithmetic mean of the first 50 positive even numbers?

(A) 50
(B) 50.5
(C) 51
(D) 51.5
(E) 52

11. Bob takes medicine every four hours. The medicine, however, is diluted by 20% every four hours. How much of the medicine remains in his body right after the 7th injection, where each intake is exactly $10ml$?

(A) 24.40
(B) 26.89
(C) 29.52
(D) 36.89
(E) 39.51

12. Bob argues that $1+2+4+\cdots$ is $-1$ because of the infinite geometric series formula insists that the sum equals $\frac{1}{1-2}$. Which of the following best explains why his argument is false?

(A) The first term should not be positive.
(B) The common ratio should not be positive.
(C) The common ratio should not be an integer.
(D) The geometric series does not work for integers.
(E) The common ratio for infinite geometric series should not be greater than 1, nor smaller than -1.

## 14.4 Matrices

Matrix is an array of numbers in rows and columns. SAT Math Level 2 test asks students about the conditions of multiplication, the meaning of determinants, and simple algebra related to matrices.

- $A_{m\times n} \times B_{n\times p} = AB_{m\times p}$.
- Given a matrix $\begin{bmatrix} a & b \\ c & d \end{bmatrix}$, the determinant is $ad - bc$, where $ad - bc = 0$ means there is no inverse matrix.
- $\begin{bmatrix} a & b \\ c & d \end{bmatrix} \begin{bmatrix} x \\ y \end{bmatrix} = \begin{bmatrix} M \\ N \end{bmatrix}$ has no solution for $x$ and $y$ if $ad - bc = 0$.
- $\begin{bmatrix} a & b \\ c & d \end{bmatrix} \pm \begin{bmatrix} p & q \\ r & s \end{bmatrix} = \begin{bmatrix} a\pm p & b\pm q \\ c\pm r & d\pm s \end{bmatrix}$, adding/subtracting entries by entries.

---

**MATHEMATICS LEVEL 2 Test - *Continued***

USE THIS SPACE FOR SCRATCH WORK.

13. If $A_{3\times 5} \times B_{m\times n} = C_{3\times 4}$, what is the value of $m \times n$?

(A) 9 (B) 12 (C) 15 (D) 20 (E) 25

14. If the following matrix equation

$$\begin{bmatrix} 5 & a \\ b & 2 \end{bmatrix} \begin{bmatrix} x \\ y \end{bmatrix} = \begin{bmatrix} 3 \\ 2 \end{bmatrix}$$

has a unique solution, which of the following values CANNOT be the value of $a \times b$?

(A) 2 (B) 3 (C) 5 (D) 7 (E) 10

**MATHEMATICS LEVEL 2 Test - *Continued***

15. Which of the following matrices has no inverse matrix?

USE THIS SPACE FOR SCRATCH WORK.

(A) $\begin{bmatrix} 1 & 3 \\ 3 & 2 \end{bmatrix}$

(B) $\begin{bmatrix} 2 & -1 \\ 3 & 11 \end{bmatrix}$

(C) $\begin{bmatrix} -3 & 1 \\ 2 & -5 \end{bmatrix}$

(D) $\begin{bmatrix} 2 & 7 \\ -1 & 4 \end{bmatrix}$

(E) $\begin{bmatrix} 1 & -2 \\ 3 & -6 \end{bmatrix}$

## Answer Key to Practice Problems

1. (D)

2. (B)

3. (B)

4. (D)

5. (B)

6. (B)

7. (D)

8. (A)

9. (D)

10. (C)

11. (E)

12. (E)

13. (D)

14. (E)

15. (E)

## Detailed Solution for Practice Problems

**1.**

- If 2 is prime, then $2(2)+1=5$ is prime. (True)
- If 3 is prime, then $2(3)+1=7$ is prime. (True)
- If 5 is prime, then $2(5)+1=11$ is prime. (True)
- If 7 is prime, then $2(7)+1=15$ is prime. (False)
- If 11 is prime, then $2(11)+1=23$ is prime. (True)

The answer must be **(D)**.

**2.**

I. $x=\frac{1}{2}$ is a counterexample.

II. True statement because its contrapositive is obviously true.

III. $x=1, y=0$ is a counterexample.

Hence, the answer is **(B)**.

**3.**

2 is prime but not odd. Therefore, the answer is **(B)**.

**4.**

$$\begin{aligned} y &= \frac{k}{\sqrt{x}} \\ 6 &= \frac{k}{\sqrt{4}} \\ 12 &= k \\ 3 &= \frac{12}{\sqrt{x}} \\ \sqrt{x} &= 4 \\ x &= 16 \end{aligned}$$

The answer is **(D)**.

**5.**

$$\begin{aligned} y &= kx^2 \\ 27 &= k(3)^2 \\ 3 &= k \\ 12 &= 3x^2 \\ 4 &= x^2 \\ 2 &= x \end{aligned}$$

Hence, the answer is **(B)**.

**6.**

Notice that an arithmetic sequence $a_n$ always satisfies the formula $a_n = a_1 + (n-1)d$. Applying it directly to the question, we get

$$\begin{cases} a_1 + 3d = 10 \\ a_1 + 6d = 16 \end{cases}$$

where $d = 2$ and $a_1 = 4$. Therefore, the answer must be **(B)**.

**7.**

A geometric sequence $b_n$ satisfies the formula $b_n = b_1(r)^{n-1}$. Applying it to the question, we get

$$\begin{cases} b_1 r^4 = 20 \\ b_1 r^6 = 5 \end{cases}$$

where $r^2 = \frac{1}{4}$. Hence, $r = \pm\frac{1}{2}$, so the answer is **(D)**.

**8.**

This is an infinite geometric series with the first term of 1 and the common ratio of $\frac{1}{2}$, so

$$\frac{1}{1-\frac{1}{2}} = 2$$

which implies that the answer is **(A)**.

**9.**

$$\begin{aligned}1+2+\cdots+100 &= \frac{100(1+100)}{2} \\ &= 50(101) \\ &= 5050\end{aligned}$$

so the answer is **(D)**.

**10.**

$$\begin{aligned}\frac{2+4+6+\cdots+100}{50} &= \frac{50(2+100)}{2(50)} \\ &= \frac{102}{2} \\ &= 51\end{aligned}$$

Thus, the answer is **(C)**.

**11.**

- The remaining amount of first intake in Bob's body is $10(0.8)^6 ml$.
- The remaining amount of second intake in Bob's body is $10(0.8)^5 ml$.
- The remaining amount of third intake in Bob's body is $10(0.8)^4 ml$.
- The remaining amount of fourth intake in Bob's body is $10(0.8)^3 ml$.
- The remaining amount of fifth intake in Bob's body is $10(0.8)^2 ml$.
- The remaining amount of sixth intake in Bob's body is $10(0.8) ml$.
- The remaining amount of seventh intake in Bob's body is $10 ml$.

Using the geometric series formula, we compute it as

$$\begin{aligned}10+10(0.8)+10(0.8)^2+\cdots+10(0.8)^6 &= \frac{10(1-(0.8)^7)}{1-0.8} \\ &\approx 39.51\end{aligned}$$

Therefore, the answer is **(E)**.

**12.**

Bob's argument is false because infinite geometric series formula can be applied for $|r| < 1$. Hence, the answer must be **(E)**.

**13.**

$A_{3\times5} \times B_{5\times4} = C_{3\times4}$, so the value of $m \times n = 20$. Hence, the answer is **(D)**.

**14.**

If the system of equation has a unique solution, then the determinant of the coefficient matrix is nonzero. Hence, $10 - ab \neq 0$. Therefore, $ab \neq 10$. The answer is **(E)**.

**15.**

- The determinant is $1(2) - 3(3) \neq 0$, so the inverse matrix exists.
- The determinant is $2(11) - (-1)(3) \neq 0$, so the inverse matrix exists.
- The determinant is $(-3)(-5) - (1)(2) \neq 0$, so the inverse matrix exists.
- The determinant is $2(4) - 7(-1) \neq 0$, so the inverse matrix exists.
- The determinant is $1(-6) - (-2)(3) = 0$, so the inverse matrix does not exist.

Hence, the answer is **(E)**.

# Topic 15

# Statistics

✓ Descriptive Statistics

✓ Box-Whisker Plots

✓ Normal Distribution

## 15.1 Descriptive Statistics

The first aspect of descriptive statistics deals with the measures of central tendency, illustrated by the following three values.

- Mean = the average of all data.
- Median = the middle value of all data arranged in increasing order.
- Mode = the most frequently appearing data.

Out of data set values, the value that is extremely off from the central measures is called an outlier. Getting rid of an outlier, we observe that mean is changed more drastically than median.

---

**MATHEMATICS LEVEL 2 Test - *Continued***

USE THIS SPACE FOR SCRATCH WORK.

1. Given a data set of $\{1,2,2,3,4,5,6\}$, which of the following is greatest of all?

(A) Mode
(B) Median
(C) Mean
(D) Minimum
(E) Lower Quartile

2. Which of the following is the outlier out of the test scores $\{750,760,800,720,580\}$?

(A) 580
(B) 720
(C) 750
(D) 760
(E) 800

**MATHEMATICS LEVEL 2 Test - *Continued***

USE THIS SPACE FOR SCRATCH WORK.

3. The mean of five positive integers is 5 where their median is 5, as well. If the list has a single mode of 3, what must be the largest possible value of all of the five integers?

(A) 5
(B) 6
(C) 7
(D) 8
(E) None of the above

4. Suppose there is a data list of $\{3,3,5,5,100\}$. If 100 is switched to 5, which of the following is least affected by this change?

(A) Mean
(B) Median
(C) Mode
(D) Range
(E) Maximum

5. If $\{a_1, a_2, a_3, \cdots, a_n\}$ has the mean value of 10, what must be true about the mean of $\{a_1+1, a_2+1, \cdots, a_n+1\}$?

(A) 10
(B) 11
(C) 12
(D) 20
(E) 21

The second aspect of it deals with the measures of spread, which includes

- Range = the difference between maximum and minimum.
- Inter-quartile Range = the difference between the upperquartile and the lowerquartile
- Upper-quartile = the median of upper half.
- Lower-quartile = the median of lower half.
- Standard Deviation = the amount of dispersion from the mean.

During the exam, it is unlikely for you to find the exact computation of standard deviation. Nevertheless, the test frequently asks you to compare the standard deviations using the following rule.

More clustered data means smaller standard deviation.

---

**MATHEMATICS LEVEL 2 Test - *Continued***

USE THIS SPACE FOR SCRATCH WORK.

6. Which of the following distributions has the smallest standard deviation?

(A) 1,2,3,4,5
(B) 0,1,3,5,6
(C) 1,3,3,3,5
(D) 1,1,3,5,5
(E) 0,2,3,4,6

7. What percent of range is the inter-quartile range for $\{21, 34, 51, 11, 13, 24, 50\}$?

(A) 25.5
(B) 50
(C) 75
(D) 87.5
(E) 92.5

## 15.2 Box-Whisker Plots

Box and Whisker plots have five important numbers that summarize the distribution of data values.

minimum $\longleftrightarrow$ lower-quartile $\longleftrightarrow$ median $\longleftrightarrow$ upper-quartile $\longleftrightarrow$ maximum

- 25% of the data lie between two consecutive five-numbers in box-whisker plots.
- It is extremely difficult to pin-point down where the mean value is from the box-whisker plots.
- The length of the box is the inter-quartile range, also known as IQR.
- The length between the endpoints of whiskers is the range.

One thing to notice from the box-whisker plots is that it does not show where the mean value is. We can guess where it could be , but it does not tell us the exact location of the standard deviation. Hence, extrapolating the standard deviation from the box-whisker plot is unwarranted, except when the box-whisker plots have overtly different sizes.

Furthermore, if the distance between any two consecutive numbers in the five number summary of box whisker plot is short, it tells that the data is dense around that short region.

---

**MATHEMATICS LEVEL 2 Test - *Continued***

8.

USE THIS SPACE FOR SCRATCH WORK.

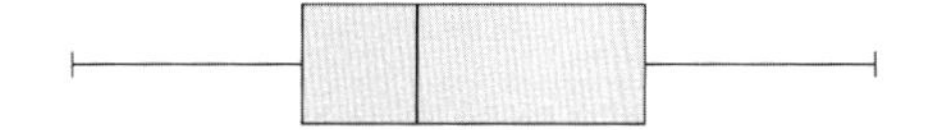

Suppose there are 100 student data in the above whisker plot, how many students are located in the region between the lower-quartile and the upper-quartile?

(A) 25
(B) 50
(C) 75
(D) 100
(E) Not enough information

**MATHEMATICS LEVEL 2 Test** - ***Continued***

USE THIS SPACE FOR SCRATCH WORK.

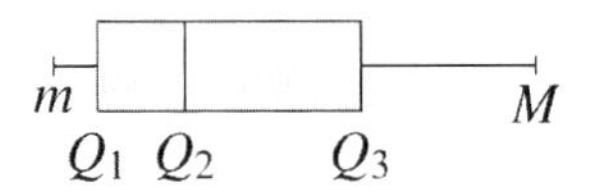


9. Given the box-whisker plot above, which of the following must be true?

(A) The inter-quartile range is half the range.
(B) Roughly 50% of the data are between $Q_2$ and $Q_3$.
(C) The range equals the standard deviation.
(D) The lower 50% of data is more clustered than the other half.
(E) Median and mode are equal to each other.

10.

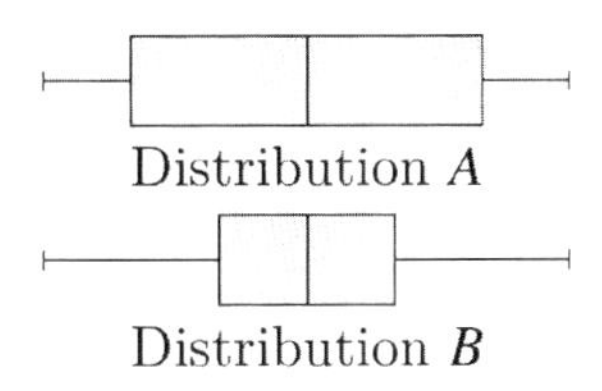


Given the two distributions with equal mean value, both distributions are symmetric about the median value. Which of the following must be true?

I. The distribution $B$ has exactly same mode.

II. The inter-quartile range of distribution $A$ is smaller than that of distribution $B$.

III. The distribution $A$ has larger standard deviation.

(A) I only
(B) II only
(C) III only
(D) I and II
(E) II and III

## 15.3 Normal Distribution

If a data set of continuous variable is normally distributed, then the following normal curve is drawn such that the highest peak is mean, median, and mode at the same time, which means that the graph is symmetric about the vertical line that passes through the mean value.

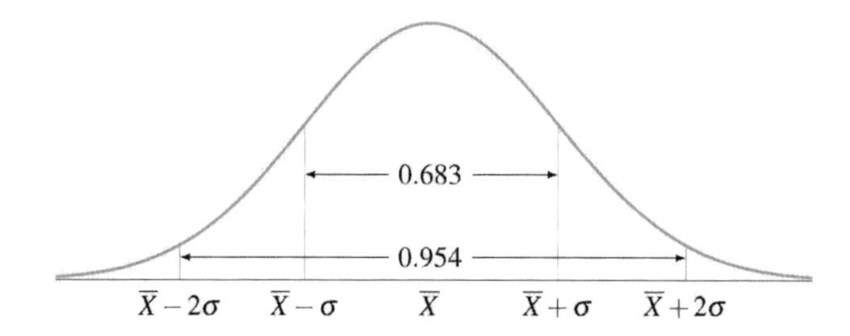


Normally, the data set does not have the mean value of 0, nor the standard deviation of 0. However, there is a technique called $Z-$ transformation that converts any specific normal distribution into the standard normal distribution with the mean of 0 and standard deviation of 1. $Z-$ transformation is specifically good for relative comparison between data values from different data sets with normal distributions.

$$Z-\text{score}=\frac{\text{data}-\text{mean}}{\text{standard deviation}}$$

Let's have a look at the graph of standard normal distribution curve.

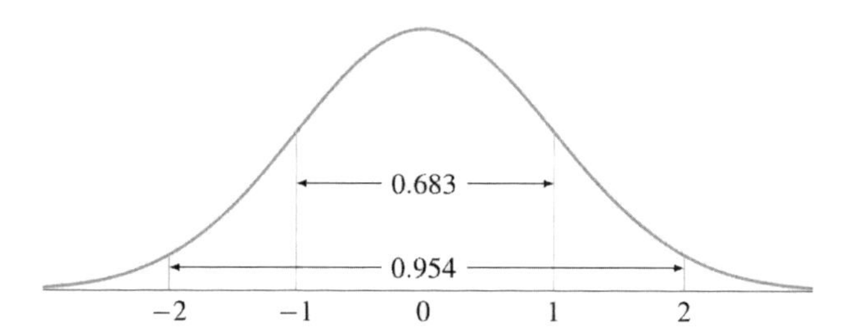


- $Z=0$ is the line that is 0 standard deviation away from the mean.
- As shown in the figure above, the standard deviation shows an interesting property that approximately 68% of the values lie within one standard deviation from the mean and about 95% of the values like within two standard deviations from the mean. Within three standard deviations, nearly all portions of the graph can be fully explained.
- The area under the curve is equal to 1, and portions of the graph mean the probability that the data is included in that region.

**MATHEMATICS LEVEL 2 Test - *Continued***

USE THIS SPACE FOR SCRATCH WORK.

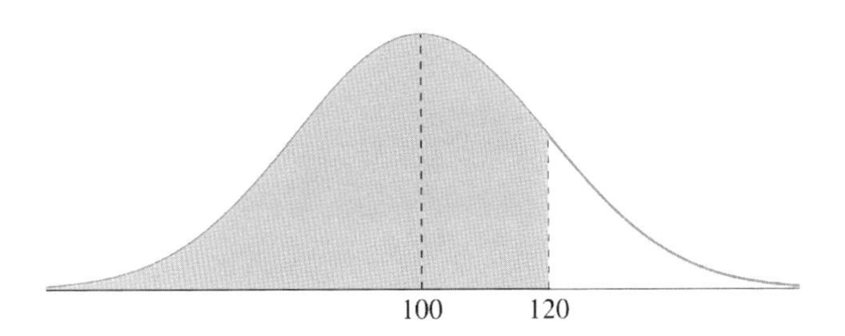


11. Given the normal distribution curve, which of the following is the $z$-value for 120 where the standard deviation is 10 and the mean value is 100?

(A) 2
(B) 1
(C) 0
(D) $-1$
(E) $-2$

12. The line of symmetry for a normal distribution curve passes through which of the following values?

(A) Maximum
(B) Minimum
(C) Mean
(D) Lower-quartile
(E) Upper-quartile

13. What is the value of standard deviation from the standard normal distribution curve?

(A) 0
(B) 1
(C) 2
(D) 3
(E) None of the above

**MATHEMATICS LEVEL 2 Test - *Continued***

14. If a data set $X$ is normally distributed, the distribution is best explained within how many standard deviations away from the mean?

USE THIS SPACE FOR SCRATCH WORK.

(A) 0
(B) 0.5
(C) 1
(D) 1.5
(E) 2

15. Assuming all three SAT Math Level 2 test scores are normally distributed, which of the following lists the scores from least to greatest, in increasing order?

| Test | Score | Average | Standard Deviation |
|---|---|---|---|
| 1 | 750 | 700 | 30 |
| 2 | 760 | 760 | 10 |
| 3 | 700 | 600 | 50 |

(A) 1, 2, and 3
(B) 2, 1, and 3
(C) 2, 3, and 1
(D) 3, 1, and 2
(E) 3, 2, and 1

## Answer Key to Practice Problems

1. (C)

2. (A)

3. (D)

4. (B)

5. (B)

6. (C)

7. (E)

8. (B)

9. (D)

10. (C)

11. (A)

12. (C)

13. (B)

14. (E)

15. (B)

## Detailed Solution for Practice Problems

**1.**

- Mode is 2.
- Median is 3.
- Mean is $\frac{1+2+2+3+4+5+6}{7} = \frac{23}{7} \approx 3.285$.
- Minimum is 1.
- Lower Quartile is 2.

Hence, the answer is **(C)**.

**2.**

The outlier is a data that is extremely off from the central measures, and in this question, it must be 580. Therefore, the answer is **(A)**.

**3.**

Let $a$ and $b$ be the unknown integers. Then, the list can be arranged as $3,3,5,a,b$. Since the mean value is 5, we compute

$$\begin{aligned} \frac{3+3+5+a+b}{5} &= 5 \\ 11+a+b &= 25 \\ a+b &= 14 \end{aligned}$$

Since $5 < a$, we get $(a,b) = (6,8),(7,7)$. If $(a,b) = (7,7)$, then the condition about single mode is breached. Hence, the largest possible value must be 8. The answer is **(D)**.

**4.**

A data list of $\{3,3,5,5,100\}$ is switched to $\{3,3,5,5,5\}$. Mean, mode, range, and maximum are changed. However, median is not changed a bit. Therefore, the answer is **(B)**.

**5.**

Adding 1 to each data of the list changes the mean value by 1. However, it does not change the standard deviation nor range. Hence, the answer is **(B)**.

**6.**

First, eliminate some answer choices by computing the range. (B) and (D) have the range of 6, so eliminate these two choices. (A), (C), and (D) have the mean value of 3. The more 3 appears in the list, the smaller the standard deviation is. The answer is **(C)**.

**7.**

Rearrange the list in increasing order : $\{11, 13, 21, 24, 34, 50, 51\}$. The interquartile range $50 - 13 = 37$, and the range is $51 - 11 = 40$. Hence, $\frac{37}{40} \times 100 = 92.5(\%)$. The answer is **(E)**.

**8.**

The box of the box-whisker plot contains 50% of the data, so the answer is **(B)**.

**9.**

50% of the total data list must be contained between $m$ and $Q_2$, whereas the other 50% between $Q_2$ and $M$. Hence, the correct answer must be **(D)** because the length between $m$ and $Q_2$ is shorter than that between $Q_2$ and $M$.

**10.**

I. Nobody knows about mode information from box-whisker plots.

II. The inter-quartile range of $A$ is longer than that of $B$.

III. The whiskers are more compact. Hence, the standard deviation of $A$ is greater than the standard deviation of $B$. In fact, according to the study of statistics, we can estimate the standard deviation as $\frac{3}{4} \times \text{IQR}$ or $\frac{1}{4} \times \text{range}$.

Hence, the answer is **(C)**.

**11.**

$$\begin{aligned} Z-\text{score} &= \frac{120-100}{10} \\ &= 2 \end{aligned}$$

Thus, the answer is **(A)**.

**12.**

The line of symmetry of a normal distribution passes through mean, median and mode. Therefore, the answer must be **(C)**.

**13.**

The standard deviation value from the standard normal distribution curve is 1. Have a look at the figure drawn in section 3. The answer is **(B)**.

**14.**

The empirical rule tells us that 68% of the data is explained within 1 standard deviation, 96% of it is explained within 2 standard deviations, and 99.7% of it is explained within 3 standard deviations. Since the answer choices have the maximum value of 2, the answer must be **(E)**.

**15.**

- Test 1 : $Z_1 = \dfrac{750-700}{30} = \dfrac{5}{3}$
- Test 2 : $Z_2 = \dfrac{760-760}{10} = 0$
- Test 3 : $Z_3 = \dfrac{700-600}{50} = 2$

$0 < \dfrac{5}{3} < 2$ implies that the score for test 3 is best, test 1 second, and test 2 worst. Therefore, the answer must be **(B)**.

## Standard Deviation

$$s = \sqrt{\frac{(x_1 - \overline{X})^2 + (x_2 - \overline{X})^2 + \cdots (x_n - \overline{X})^2}{n}}$$

In SAT Math Level 2, there are few questions related to calculating the standard deviation. However, the formula does tell us what to look for. First, we need the mean value $\overline{X}$, which is computed by

$$\overline{X} = \frac{x_1 + x_2 + \cdots + x_n}{n}$$

The standard deviation measures how much data values deviate from the mean value. The reason why we square the difference between each data value and the mean is because differences MAY cancel one another. Hence, we add squares of the differences, which are fairly large by itself. Therefore, we take a square root of the number to reduce the size of the number.

From the box-whisker plot, there are two possible approximations of standard deviation, without figuring out the exact value of the mean value.

- The standard deviation is approximately equal to $\frac{\text{range}}{4}$.
- The standard deviation is approximately equal to $\frac{3}{4} \times \text{IQR}$.

As they are approximate values, use them to get the rough idea of what standard deviation might be.

# Minitest 1

**Minitest 1** includes

- System of Equations
- Laws of Sines and Laws of Cosines
- Factorials
- Space Geometry
- Sequence and Series
- Limits
- Polynomial Functions and Equations
- Factor(and Remainder) Theorem
- Even and Odd Functions
- Inequalities
- Conic Sections
- Quadratic Functions and Equations
- Number Theory
- Trigonometric Ratio

# MATHEMATICS LEVEL 2 MINITEST 1

REFERENCE INFORMATION

THE FOLLOWING INFORMATION IS FOR YOUR REFERERENCE IN ANSWERING SOME OF THE QUESTIONS IN THE TEST.

Volume of a right circular cone with radius $r$ and height $h$ : $V = \frac{1}{3}\pi r^2 h$

Volume of a sphere with radius $r$ : $V = \frac{4}{3}\pi r^3$

Volume of a pyramid with base area $B$ and height $h$ : $V = \frac{1}{3}Bh$

Surface Area of a sphere with radius $r$ : $S = 4\pi r^2$

DO NOT DETACH FROM THE BOOK.

## MATHEMATICS LEVEL 2 MINITEST 1

For each of the following problems, decide which is the BEST of the choices given. If the exact numerical value is not one of the choices, select the choice that best approximates this value. Then fill in the corresponding circle on the answer sheet.

Notes: (1) A scientific or graphing calculator will be necessary for answering some (but not all) of the questions in this test. For each question you will have to decide whether or not you should use a calculator.

(2) For some questions in this test you may have to decide whether your calculator should be in the radian mode or the degree mode.

(3) Figures that accompany problems in this test are intended to provide information useful in solving the problems. They are drawn as accurately as possible EXCEPT when it is stated in a specific problem that its figure is not drawn to scale. All figures lie in a plane unless otherwise indicated.

(4) Unless otherwise specified, the domain of any function $f$ is assumed to be the set of all real numbers $x$ for which $f(x)$ is a real number. The range of $f$ is assumed to be the set of all real numbers $f(x)$, where $x$ is in the domain of $f$.

(5) Reference information that may be useful in answering the questions in this test can be found on the page preceding Question 1.

1. USE THIS SPACE FOR SCRATCH WORK.

$$3x+2y=2$$
$$4x-3y=14$$

If $(x,y)$ is the solution to the system of equations above, what is the value of $x$?

(A) $-2$
(B) 4
(C) 3
(D) 2
(E) 0

**MATHEMATICS LEVEL 2 Test - *Continued***

USE THIS SPACE FOR SCRATCH WORK.

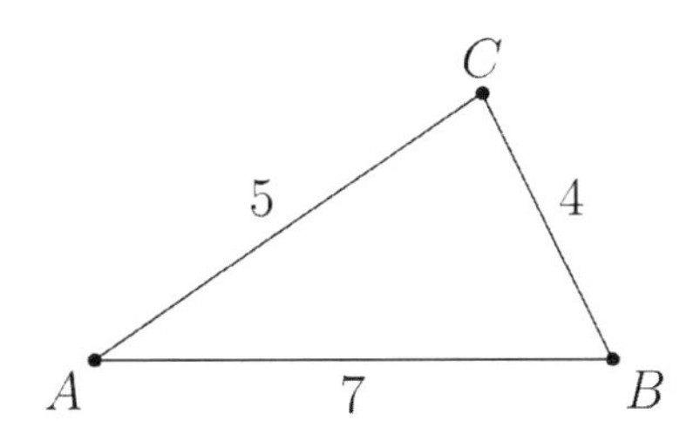


2. Given a triangle *ABC* with the following side lengths of 4,5, and 7, then the area of the triangle *ABC* equals

(A) 9.80
(B) 9.92
(C) 10.2
(D) 11.43
(E) 11.56

3. If $\frac{(n+1)!}{(n-1)!} = 56$, what is the value of $n$?

(A) 6
(B) 7
(C) 8
(D) 10
(E) 12

4. The radius of the base of a right circular cone is 5, and the slant height of the cone is 10. What is the surface area of the cone?

(A) $50\pi$
(B) $68\pi$
(C) $70\pi$
(D) $75\pi$
(E) $80\pi$

**MATHEMATICS LEVEL 2 Test - *Continued***

USE THIS SPACE FOR SCRATCH WORK.

5. If the 11th term of an arithmetic sequence is 30 and the 21st term is 0, what is the 10th term of the sequence?

(A) 30
(B) 33
(C) 36
(D) 39
(E) 42

6. $\lim_{n\to\infty} \dfrac{1}{\sqrt{n^2+n}-n} =$

(A) 0
(B) 1
(C) 2
(D) 10
(E) infinity

7. If one of the roots of a polynomial function $f(x)=0$ is $1+\dfrac{\sqrt{3}}{2}i$, which of the following could be $f(x)$?

(A) $2x^2-4x+6$
(B) $x^2-2x+10$
(C) $3x^2+6x-4$
(D) $4x^2-8x+7$
(E) $5x^2-10x+7$

**MATHEMATICS LEVEL 2 Test - *Continued***

8. If $p(x) = x^3 - 2x^2 + mx + n$ is divisible by $x^2 - 3x + 2$, what is the value of $m$?

USE THIS SPACE FOR SCRATCH WORK.

(A) $-2$
(B) $-1$
(C) 0
(D) 1
(E) 2

9. If $f(x) = \cos(x)$ and $g(x) = x^2 + 3$, which of the following is not true?

(A) $f(x) \cdot g(x) + 2$ is an even function.
(B) $f(x) + g(x)$ is an even function.
(C) $g(f(x))$ is an even function.
(D) $f(x) - g(x) + 2$ is an even function.
(E) $f(x-1) + g(x-1)$ is an even function.

10. Which of the following conditions satisfies $\dfrac{x}{(x-1)(x+2)} < 0$?

(A) $x < 0$ or $x > 2$
(B) $x < -2$ or $x > 1$
(C) $x < 1$ or $x > 0$
(D) $x < -2$ or $0 < x < 1$
(E) $-2 < x < 0$ or $x > 1$

**MATHEMATICS LEVEL 2 Test - *Continued***

USE THIS SPACE FOR SCRATCH WORK.

11. Given the hyperbola equation $16x^2 - 25y^2 = 400$, which of the following must be the intersection point of the asymptotes?

(A) $(4,5)$
(B) $(5,4)$
(C) $(0,0)$
(D) $(-4,5)$
(E) $(4,-5)$

12. If $kx^2 + 4x - 2 = 0$ have two different real roots, which of the following must be true?

(A) $k > -2$
(B) $k < -2$
(C) $-2 < k < 0$ or $k > 0$
(D) $-2 < k < 2$
(E) $-1 < k < 3$

13. If $x^{101} + 4x^{100}$ is a perfect square, where $x$ is an integer, which of the following could be the value of $x$?

(A) 1
(B) 4
(C) 10
(D) 13
(E) 21

**MATHEMATICS LEVEL 2 Test - *Continued***

USE THIS SPACE FOR SCRATCH WORK.

14. The arithmetic mean of three distinct positive integers is 38, and the largest number of these three is 43. Of the remaining positive integers, which of the following could be the LEAST number?

(A) 19
(B) 25
(C) 30
(D) 40
(E) 71

15. If $\sin(x) = t$ for $0 < x < \frac{\pi}{2}$, then $\tan(x) =$

(A) $\frac{t}{\sqrt{t^2-1}}$
(B) $\frac{t}{\sqrt{1-t^2}}$
(C) $\frac{1}{\sqrt{1-t^2}}$
(D) $\frac{t}{\sqrt{t^2+1}}$
(E) $\frac{t}{\sqrt{t^2+1}}$

## Answerkey to Minitest 1

1. (D)

2. (A)

3. (B)

4. (D)

5. (B)

6. (C)

7. (D)

8. (B)

9. (E)

10. (D)

11. (C)

12. (A)

13. (E)

14. (C)

15. (B)

## Solution

1. Eliminate $y$ to get $17x = 34$, so $x = 2$.

2. Apply Heron's formula. $\sqrt{s(s-4)(s-5)(s-7)}$ where $s = \dfrac{4+5+7}{2} = 8$.

3. Simplify the left-hand side to get $(n+1)n = 56$, so $n = 7$.

4. The surface area of the cone is $\pi r^2 + \pi rl = 25\pi + 50\pi = 75\pi$.

5. Let $a_{11} = a_1 + 10d = 30$ and $a_{21} = a_1 + 20d = 0$, so, $a_1 = 60$ and $d = -3$. Hence, $a_{10} - 3 = a_{11}$, so $a_{10} = 33$.

6. Write the expression in your calculator to get 2, or multiply $\sqrt{n^2+n} + n$ to both numerator and denominator to get 2.

7. $x = 1 + \dfrac{\sqrt{3}}{2} \rightarrow x - 1 = \dfrac{\sqrt{3}}{2}i \rightarrow (x-1)^2 = -\dfrac{3}{4} \rightarrow -4(x-1)^2 = 3 \rightarrow -4(x^2 - 2x + 1) = 3 \rightarrow$
$-4x^2 + 8x - 4 - 3 = 0 \rightarrow -4x^2 + 8x - 7 = 0 \rightarrow 4x^2 - 8x + 7 = 0$

8. According to factor theorem, $p(1) = 1^3 - 2(1)^2 + m + n = 0$ and $p(2) = 2^3 - 2(2)^2 + 2m + n = 0$, so $m + n = 1$ and $2m + n = 0$. Therefore, $m = -1$.

9. Translating right by 1 shifts the line of symmetry to $x = 1$, so the answer is **(E)**.

10. $\dfrac{x}{(x-1)(x+2)} < 0 \rightarrow x(x-1)(x+2) < 0 \rightarrow x < -2$ or $0 < x < 1$.

11. $16x^2 - 25y^2 = 400 \rightarrow \dfrac{x^2}{25} - \dfrac{y^2}{16} = 1$, so $y = \pm\dfrac{4}{5}x$ are the asymptotes of the hyperbola.

12. $kx^2 + 4x - 2 = 0$ has two real roots if $D = 4^2 - 4(-2)k > 0 \rightarrow 8k > -16 \rightarrow k > -2$.

13. $x^{101} + 4x^{100} = x^{100}(x+4)$ is a perfect square if $x + 4$ is a perfect square.

14. Let $a$, $b$, and 43 be the three numbers. Then, $\dfrac{a+b+43}{3} = 38$, so $a + b + 43 = 114$. Hence, $a + b = 71$. Substituting answers, 30 is the least out of all answer choices.

15. If $\sin(x) = \dfrac{t}{1}$, then the adjacent leg has the length of $\sqrt{1-t^2}$ by Pythagorean Theorem. Hence, $\tan(x) = \dfrac{t}{\sqrt{1-t^2}}$.

# Minitest 2

**Minitest 2** includes

- Laws of Sines and Cosines
- Trigonometric Identities
- Linear Functions and Equations
- Sequence and Series
- Polynomial Functions and Equations
- Trigonometric Functions
- Periodic Functions
- Limits
- Plane Geometry
- System of Equations
- Even and Odd Functions
- Space Geometry
- Logarithm
- Vectors

# MATHEMATICS LEVEL 2 MINITEST 2

REFERENCE INFORMATION

THE FOLLOWING INFORMATION IS FOR YOUR REFERERENCE IN ANSWERING SOME OF THE QUESTIONS IN THE TEST.

Volume of a right circular cone with radius $r$ and height $h$ : $V = \frac{1}{3}\pi r^2 h$

Volume of a sphere with radius $r$ : $V = \frac{4}{3}\pi r^3$

Volume of a pyramid with base area $B$ and height $h$ : $V = \frac{1}{3}Bh$

Surface Area of a sphere with radius $r$ : $S = 4\pi r^2$

DO NOT DETACH FROM THE BOOK.

## MATHEMATICS LEVEL 2 MINITEST 2

For each of the following problems, decide which is the BEST of the choices given. If the exact numerical value is not one of the choices, select the choice that best approximates this value. Then fill in the corresponding circle on the answer sheet.

Notes: (1) A scientific or graphing calculator will be necessary for answering some (but not all) of the questions in this test. For each question you will have to decide whether or not you should use a calculator.

(2) For some questions in this test you may have to decide whether your calculator should be in the radian mode or the degree mode.

(3) Figures that accompany problems in this test are intended to provide information useful in solving the problems. They are drawn as accurately as possible EXCEPT when it is stated in a specific problem that its figure is not drawn to scale. All figures lie in a plane unless otherwise indicated.

(4) Unless otherwise specified, the domain of any function $f$ is assumed to be the set of all real numbers $x$ for which $f(x)$ is a real number. The range of $f$ is assumed to be the set of all real numbers $f(x)$, where $x$ is in the domain of $f$.

(5) Reference information that may be useful in answering the questions in this test can be found on the page preceding Question 1.

USE THIS SPACE FOR SCRATCH WORK.

1. There is a cube with the side length of 10. The area of triangle $ABC$ equals

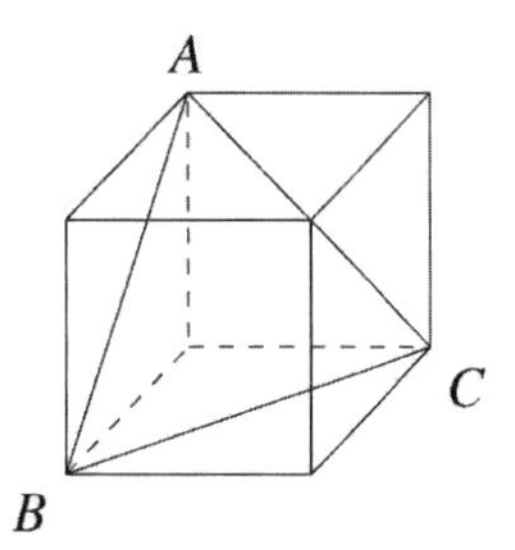


(A) 129.90
(B) 88.83
(C) 86.60
(D) 82.37
(E) 50.00

**MATHEMATICS LEVEL 2 Test -** ***Continued***

2. If $\theta$ is the angle between the positive $x$-axis, and the radius of the unit circle, centered at the origin with the length of 1, then
$\sin(90° - \theta) + \cos(180° - \theta) - \tan(180° - \theta) =$

USE THIS SPACE FOR SCRATCH WORK.

(A) $\sin\theta$
(B) $\cos\theta$
(C) $\tan\theta$
(D) $\sin\theta - \cos\theta$
(E) $\cos\theta - \sin\theta$

3. If $f(2x-3) = 4x-2$, then $f(x) =$

(A) $x+4$
(B) $x-4$
(C) $2x+4$
(D) $3x+5$
(E) $3x-5$

4. $x, 12, 3x-6, \cdots$ are the first three terms in a geometric sequence whose terms are always positive. Then, the 5th term of the sequence equals

(A) 1.5
(B) 36.5
(C) 40.5
(D) 60
(E) 96

**MATHEMATICS LEVEL 2 Test - *Continued***

5. Which of the following could be true for $p(x) = ax^3 + bx^2 + cx + d$, given the graph of the function?

USE THIS SPACE FOR SCRATCH WORK.

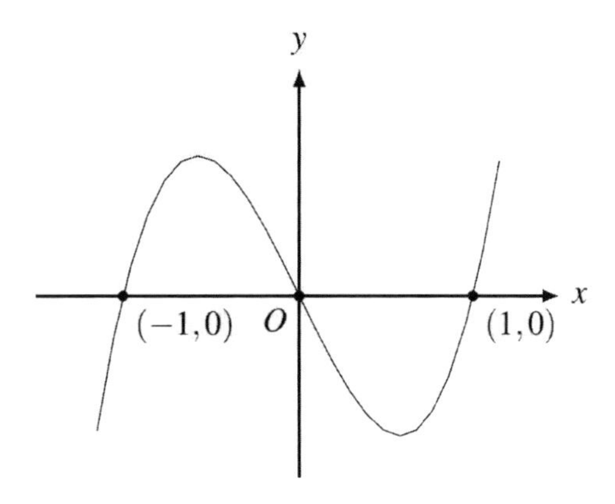


(A) $b > 0$
(B) $a < 0$
(C) $b = 0$
(D) $b < 0$
(E) $d > 0$

6. What is the period of the graph of $f(x) = 2\sin(3\pi x - \pi) + 1$?

(A) $\frac{3\pi}{2}$
(B) $\frac{2\pi}{3}$
(C) $\frac{-\pi}{2}$
(D) $\frac{\pi}{3}$
(E) $\frac{\pi}{2}$

## MATHEMATICS LEVEL 2 Test - *Continued*

USE THIS SPACE FOR SCRATCH WORK.

7. If a function $y = f(x)$ satisfies the property such that $f(x) = f(x+2)$, which of the following could be $f(x)$?

(A) $2\sin(x) + 1$

(B) $\sin(2x)$

(C) $\cos(2x) - 1$

(D) $3\tan(\pi x)$

(E) $4\tan(\frac{\pi}{2}x)$

8. $\lim_{h \to 0} \frac{h}{\sqrt{4+h} - 2} =$

(A) 0
(B) 2
(C) 4
(D) 6
(E) 8

9. If $BD = 10$, what is the length of $\overline{AC}$?

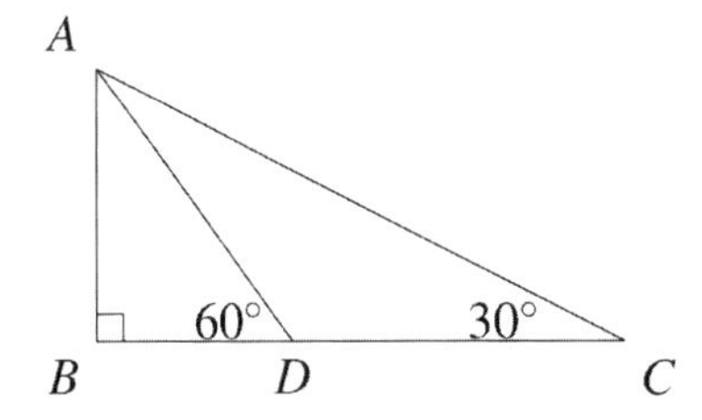


(A) 20
(B) 25.12
(C) 34.64
(D) 36
(E) 36.56

**MATHEMATICS LEVEL 2 Test - *Continued***

USE THIS SPACE FOR SCRATCH WORK.

10. Which of the following is the equation of the line that contains the common intersection points between the circles $x^2+y^2=16$ and $x^2+y^2-8x-8y+16=0$?

(A) $y=x$
(B) $y=-x$
(C) $y=x+4$
(D) $y=x-4$
(E) $y=-x+4$

11. Which of the following functions satisfies $f(-x)=-f(x)$?

I. $f(x)=x^4+5x^2-3$

II. $f(x)=3x^3+5x+1$

III. $f(x)=x^3+x$

(A) I only
(B) II only
(C) III only
(D) II and III only
(E) I, II, and III only

**MATHEMATICS LEVEL 2 Test - *Continued***

USE THIS SPACE FOR SCRATCH WORK.

12. Given a pyramid drawn below such that $\overline{AB} \perp \overline{OA} \perp \overline{OC}$, where $OA = 4$ and $OB = 3$, and $OC = 4$, the area of $\triangle ABC$ is equal to

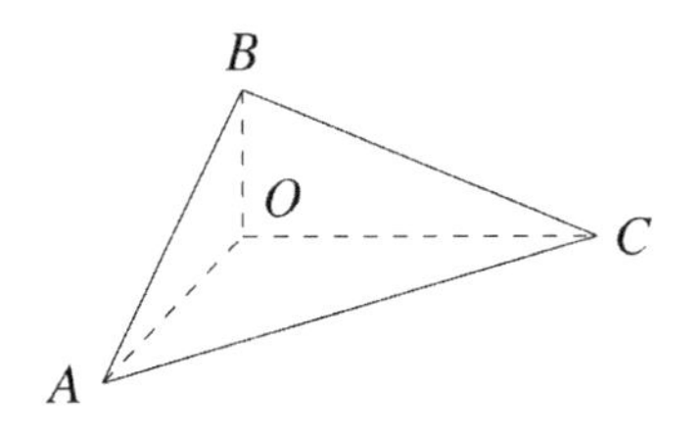


(A) 11.66
(B) 16.42
(C) 18.44
(D) 20.25
(E) 21.32

13. Bob notices that $f(x) = 3^x$ and $f^{-1}(x) = \log_3(x)$ are inverse functions to one another, learning that $f(f^{-1}(x)) = x$. Taking the practice test for SAT Math Level 2, Bob simplifies $3^{\log_9 18}$ into

(A) $3\sqrt{2}$
(B) $3\sqrt{3}$
(C) $2\sqrt{5}$
(D) $2\sqrt{6}$
(E) $2\sqrt{7}$

**MATHEMATICS LEVEL 2 Test - *Continued***

USE THIS SPACE FOR SCRATCH WORK.

14. If $\sin(x) = t$ for all $x$ in the interval $\frac{\pi}{2} < x < \pi$, then $\sin(2x) =$

(A) $\frac{t}{\sqrt{1-t^2}}$

(B) $\frac{t^2}{1-t^2}$

(C) $\frac{1}{t\sqrt{1-t}}$

(D) $-2t\sqrt{1-t^2}$

(E) $2t\sqrt{t^2-1}$

15. Vector is a mathematical object with both direction and magnitude. For example, a vector **a** is equal to $< 5, -12 >$, meaning that a point that starts off from the origin ends at the terminal point of $(5, -12)$. Unit vector is the vector in the same direction, except that the length is 1. Which of the following must be the unit vector of **a**?

(A) $(0.12, -0.38)$
(B) $(0.25, -0.37)$
(C) $(0.38, -0.92)$
(D) $(0.5, -1)$
(E) $(0.25, -0.75)$

## Answerkey to Minitest 2

1. (C)

2. (C)

3. (C)

4. (C)

5. (C)

6. (C)

7. (E)

8. (C)

9. (C)

10. (E)

11. (C)

12. (A)

13. (A)

14. (D)

15. (C)

## Solution

1. Let $AB = BC = CA = 10\sqrt{2}$. Find the height of the triangle by $AB\sin(60^\circ) = 5\sqrt{6}$. Then, the area of $ABC$ equals $\frac{1}{2} \times 10\sqrt{2} \times 5\sqrt{6} \approx 86.6$.

2. $\sin(90^\circ - \theta) + \cos(180^\circ - \theta) - \tan(180^\circ - \theta) = \cos(\theta) - \cos(\theta) - (-\tan(\theta)) = \tan(\theta)$.

3. Let $t = 2x - 3$, so $\frac{t+3}{2} = x$, Therefore, $f(t) = 4(\frac{t+3}{2}) - 2 = 2(t+3) - 2 = 2t + 4$, so $f(x) = 2x + 4$.

4. $\frac{12}{x} = \frac{3x-6}{12}$, so $3x^2 - 6x = 144$, so $x = 8$ or $-6$. In this question, $x = 8$ because all answer choices are positive. Therefore, the ratio $r$ is 1.5, so $8(1.5)^4 = 40.5$.

5. All the $x$-intercepts are $-1$, 0, and 1, so $-\frac{b}{a} = 0$, by Vieta's formula. Hence, $b = 0$.

6. $y = 2\sin(3\pi(x - \frac{1}{3})) + 1$ has the period of $\frac{2\pi}{3\pi} = \frac{2}{3}$.

7. If $f(x+2) = f(x)$, then the function $f(x)$ is periodic with the period of 2. The only function that has the period of 2 is $4\tan(\frac{\pi}{2}x)$.

8. $\lim_{h\to 0} \frac{h(\sqrt{4+h}+2)}{(\sqrt{4+h}-2)(\sqrt{4+h}+2)} = \lim_{h\to 0} \frac{h(\sqrt{4+h}+2)}{4+h-4} = \lim_{h\to 0}(\sqrt{4+h}+2) = \sqrt{4}+2 = 4$.

9. If $BD = 10$, then $AB = 10\sqrt{3}$ and $AD = 20$, by $30^\circ - 60^\circ - 90^\circ$ special right triangle ratio. Then, $AC = 2 \times AB = 2 \times (10\sqrt{3}) = 20\sqrt{3}$ by the same right triangle ratio.

10. Since $x^2 + y^2 = 16$, $x^2 + y^2 - 8x - 8y + 16 = 0$ turns into $16 - 8x - 8y + 16 = 0$, so $8x + 8y = 32$. Hence, $x + y = 4$.

11. III has the odd exponents for each term, so III is the only odd function.

12. Pythagorean theorem states that $AB = BC = 5$, and $AC = 4\sqrt{2}$. Hence, the triangle $ABC$ has the area of $\sqrt{s(s-5)(s-5)(s-4\sqrt{2})}$ where $s = \frac{5+5+4\sqrt{2}}{2}$, by Heron's formula. The expression is approximately equal to 11.66.

13. $3^{\log_9 18} = 3^{\log_{3^2} 18} = 3^{\frac{1}{2}\log_3 18} = 3^{\log_3 \sqrt{18}} = \sqrt{18} = 3\sqrt{2}$.

14. Since $\sin(2x) = 2\sin(x)\cos(x)$, and $\cos(x) = -\sqrt{1-\sin^2(x)} = -\sqrt{1-t^2}$ by C.A.S.T rule, $\sin(2x) = -2t\sqrt{1-t^2}$.

15. The unit vector of $\mathbf{a}$ is $\frac{1}{\sqrt{5^2+(-12)^2}}(5,-12) = \frac{1}{13}(5,-12) = (\frac{5}{13}, -\frac{12}{13})$.

## Heron's Formula

$$\text{Area} = \sqrt{p(p-a)(p-b)(p-c)} \text{ where } p = \frac{a+b+c}{2}$$

Heron's formula is useful to find the area of triangle when the side-lengths are all given.

$$\text{Area} = \sqrt{p(p-a)(p-b)(p-c)}$$

where $p = \frac{a+b+c}{2}$. Heron's formula can be directly deduced by the law of cosine and the law of sine. The deduction of Heron's formula requires bits of algebra.

$$\begin{aligned}
\text{Area of } \triangle\ ABC &= \frac{ab}{2}\sin(\theta) \\
&= \frac{1}{2}ab\sqrt{1-\cos^2(\theta)} \\
&= \frac{1}{2}ab\sqrt{1-\frac{(c^2-a^2-b^2)^2}{4a^2b^2}} \\
&= \sqrt{\frac{4a^2b^2-(c^2-a^2-b^2)^2}{16}} \\
&= \sqrt{\frac{(2ab-c^2+a^2+b^2)(2ab+c^2-a^2-b^2)}{16}} \\
&= \sqrt{\frac{((a+b)^2-c^2)(c^2-(a-b)^2)}{16}} \\
&= \sqrt{\frac{(a+b+c)(-a+b+c)(a-b+c)(a+b-c)}{16}} \\
&= \sqrt{p(p-a)(p-b)(p-c)}
\end{aligned}$$

# Minitest 3

**Minitest 3** includes

- Linear Functions and Equations
- Matrices
- Space Geometry
- Combinations and Permutations
- Radical Functions and Equations
- Quadratic Functions and Equations
- Trigonometric Identities
- Plane Geometry
- Sequence and Series
- Inverse Functions
- Vectors
- Parametric Functions
- Conic Sections

# MATHEMATICS LEVEL 2 MINITEST 3

REFERENCE INFORMATION

THE FOLLOWING INFORMATION IS FOR YOUR REFERERENCE IN ANSWERING SOME OF THE QUESTIONS IN THE TEST.

Volume of a right circular cone with radius $r$ and height $h$ : $V = \frac{1}{3}\pi r^2 h$

Volume of a sphere with radius $r$ : $V = \frac{4}{3}\pi r^3$

Volume of a pyramid with base area $B$ and height $h$ : $V = \frac{1}{3}Bh$

Surface Area of a sphere with radius $r$ : $S = 4\pi r^2$

DO NOT DETACH FROM THE BOOK.

## MATHEMATICS LEVEL 2 MINITEST 3

For each of the following problems, decide which is the BEST of the choices given. If the exact numerical value is not one of the choices, select the choice that best approximates this value. Then fill in the corresponding circle on the answer sheet.

Notes: (1) A scientific or graphing calculator will be necessary for answering some (but not all) of the questions in this test. For each question you will have to decide whether or not you should use a calculator.

(2) For some questions in this test you may have to decide whether your calculator should be in the radian mode or the degree mode.

(3) Figures that accompany problems in this test are intended to provide information useful in solving the problems. They are drawn as accurately as possible EXCEPT when it is stated in a specific problem that its figure is not drawn to scale. All figures lie in a plane unless otherwise indicated.

(4) Unless otherwise specified, the domain of any function $f$ is assumed to be the set of all real numbers $x$ for which $f(x)$ is a real number. The range of $f$ is assumed to be the set of all real numbers $f(x)$, where $x$ is in the domain of $f$.

(5) Reference information that may be useful in answering the questions in this test can be found on the page preceding Question 1.

1. If $\overline{OH}$ is perpendicular to the line $3x-4y=28$, then the length of $\overline{OH}$ is equal to

USE THIS SPACE FOR SCRATCH WORK.

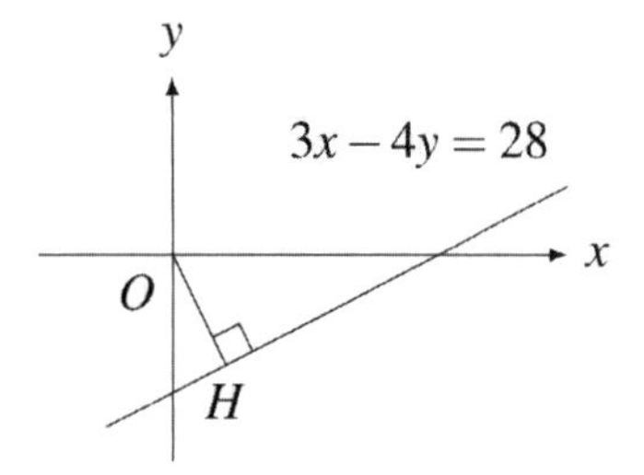


(A) 5.6
(B) 6.5
(C) 8.0
(D) 8.5
(E) 8.7

**MATHEMATICS LEVEL 2 Test - *Continued***

USE THIS SPACE FOR SCRATCH WORK.

2. If there are three matrices $A$, $B$, and $C$, such that the matrix $A$ has dimension $m \times n$, matrix $B$ has dimension $p \times m$, and matrix $C$ has dimension $n \times p$, which of the following must be true?

(A) The product $AB$ exists.
(B) The product $BC$ exists.
(C) The product $ABC$ exists.
(D) The product $CBA$ exists.
(E) The product $BCA$ exists.

3. If the height of a cylinder is increased by 10 percent, by what percent must the radius of the circular base be increased so that the volume of the cylinder is increased by 25 percent?

(A) 5.6%
(B) 6.2%
(C) 6.6%
(D) 7.5%
(E) 7.7%

4. If $\frac{n!}{4!(n-4)!} = \frac{n!}{5!(n-5)!}$, which of the following must be the value of $n$?

(A) 1
(B) 4
(C) 5
(D) 9
(E) 20

## MATHEMATICS LEVEL 2 Test - *Continued*

USE THIS SPACE FOR SCRATCH WORK.

5. If $\sqrt{x^2} = |x|$, then the solution set of $x$ consists of

(A) zero only
(B) positive real numbers only
(C) negative real numbers only
(D) all real numbers
(E) no real numbers

6. If $-1$ is a root of the equation $kx^2 + 6x - 4$, then the other root is

(A) 4  (B) 2  (C) 0.4  (D) $-0.4$  (E) $-2.5$

7. The following figure shows the graph of the equation $2x - 3y + 9 = 0$. Which of the following could be the value of $\theta$?

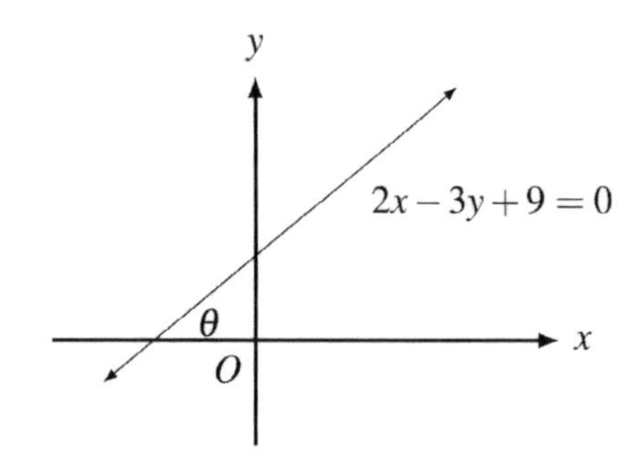


(A) 33.7°
(B) 34.2°
(C) 37.8°
(D) 38.1°
(E) 40.6°

**MATHEMATICS LEVEL 2 Test - *Continued***

USE THIS SPACE FOR SCRATCH WORK.

8. The period of the graph of $y = 3\cos^2(2x)$ equals

(A) $\frac{\pi}{8}$

(B) $\frac{\pi}{4}$

(C) $\frac{\pi}{2}$

(D) $\pi$

(E) $2\pi$

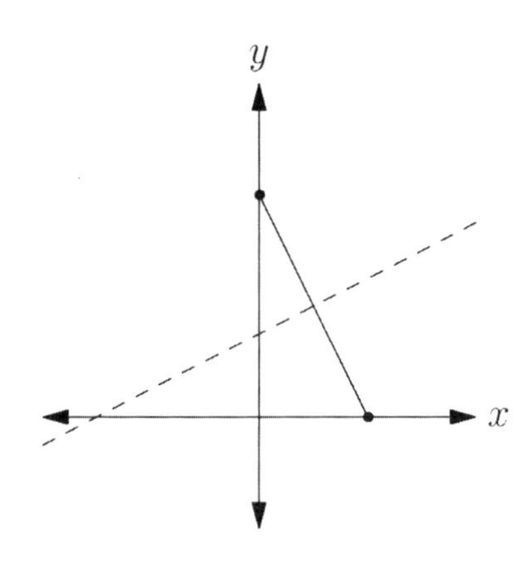


9. Which of the following is the equation whose graph is the set of points equidistant from the point (0,4) and (2,0)?

(A) $y = x + 1$

(B) $y = 2x + 1$

(C) $y = \frac{1}{2}x + \frac{1}{2}$

(D) $y = \frac{1}{2}x + \frac{3}{2}$

(E) $y = \frac{1}{4}x - \frac{3}{2}$

**MATHEMATICS LEVEL 2 Test - *Continued***

USE THIS SPACE FOR SCRATCH WORK.

10. If $a_1 = 1, a_2 = 3$ and $a_{n+1} = \dfrac{a_n + a_{n+2}}{2}$, then what is the 20th term of the sequence?

(A) 36
(B) 39
(C) 41
(D) 43
(E) 60

11. If $f(x) = x^2 + 1$, where $x \geq 0$, then $(f \circ f^{-1})(x)$ could equal?

(A) 1
(B) $x$
(C) $x^2$
(D) $x^2 + 1$
(E) $\dfrac{1}{x^2+1}$

12. Given two vectors **a** and **b**, where $|\mathbf{a}| = 10$ and $|\mathbf{b}| = 18$, then which of the following can NOT be the value of $|\mathbf{a} + \mathbf{b}|$?

(A) 7
(B) 8
(C) 15
(D) 22
(E) 28

## MATHEMATICS LEVEL 2 Test - *Continued*

USE THIS SPACE FOR SCRATCH WORK.

13. Given the parametric equations $x = \sec\theta$ and $y = \tan\theta$, which of the following is the graph of the points $(x, y)$?

(A) circle
(B) ellipse
(C) parabola
(D) hyperbola
(E) line

14. If the line $y = x + k$ is tangent to the circle $x^2 + y^2 = 4$, then the value of $k$ could be

(A) $\sqrt{2}$ (B) $\sqrt{3}$ (C) $\sqrt{5}$ (D) $2\sqrt{2}$ (E) $\sqrt{7}$

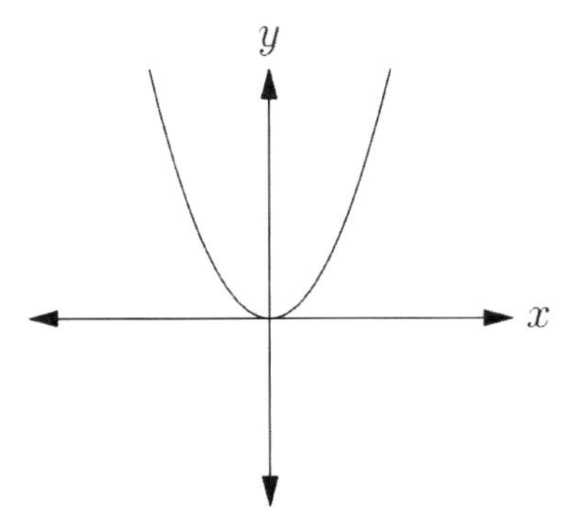


Figure of $y = 2x^2$

15. If the equation of a parabola is $y = 2x^2$, then the directrix of the graph must be $y =$

(A) 2 (B) $\frac{1}{2}$ (C) $\frac{1}{8}$ (D) $-\frac{1}{8}$ (E) $-\frac{1}{2}$

## Answerkey to Minitest 3

1. (A)
2. (D)
3. (C)
4. (D)
5. (D)
6. (C)
7. (A)
8. (C)
9. (D)
10. (B)
11. (B)
12. (A)
13. (D)
14. (D)
15. (D)

## Solution

1. The distance between (0,0) and $3x-4y-28=0$ equals $\dfrac{|3(0)-4(0)-28|}{\sqrt{3^2+(-4)^2}}=\dfrac{28}{5}=5.6$.

2. $C_{n\times p}\times B_{p\times m}\times A_{m\times n}=CBA_{n\times n}$.

3. $1.25V=\pi(\Box r)^2(1.1h)=\Box^2(1.1)\pi r^2h$ implies $\dfrac{1.25}{1.1}=\Box^2$, so $\Box=\sqrt{\dfrac{1.25}{1.1}}\approx 1.066$, meaning that 6.6% increase from the original base length.

4. $4!(n-4)!=5!(n-5)!$ holds if $n=9$. In fact, $\dbinom{n}{4}=\dbinom{n}{5}$ holds if $n=9$.

5. $\sqrt{x^2}=|x|$ for all real $x$.

6. $k(-1)^2+6(-1)-4=0$ so $k=10$. Hence, $10x^2+6x-4=(x+1)(10x-4)=0$. The root other than $-1$ is $\dfrac{4}{10}$.

7. $\tan(\theta)=\dfrac{2}{3}$, the slope of the line. Hence, $\theta=\tan^{-1}(\dfrac{2}{3})\approx 33.7°$.

8. $y=3\cos^2(2x)=3\left(\dfrac{1+\cos(4x)}{2}\right)=\dfrac{3}{2}+\dfrac{3\cos(4x)}{2}$ has the period of $\dfrac{2\pi}{4}=\dfrac{\pi}{2}$.

9. $\sqrt{(x-0)^2+(y-4)^2}=\sqrt{(x-2)^2+(y-0)^2}$ implies $x^2+y^2-8y+16=x^2-4x+4+y^2$, so $-8y+16=-4x+4$. Hence, $8y=4x+12$, so $y=\dfrac{1}{2}x+\dfrac{3}{2}$.

10. $a_{n+1}=\dfrac{a_n+a_{n+2}}{2}$ implies that $a_n$ is an arithmetic sequence, so $a_n=a_1+(n-1)d$ where $d=a_2-a_1=2$. Hence, $a_{20}=a_1+19d=1+19(2)=39$.

11. $f(f^{-1}(x))=x$, so $(f\circ f^{-1})(x)=x$.

12. $18-10\le|\mathbf{a}+\mathbf{b}|\le 18+10$, so 7 cannot be its length.

13. $x^2-y^2=\sec^2(\theta)-\tan^2(\theta)=1$, so it must be hyperbola.

14. Substituting $y=x+k$ into $x^2+y^2=4$, we get $x^2+x^2+2kx+k^2=4$, so $2x^2+2kx+(k^2-4)=0$. Its discriminant should be 0, so
$(2k)^2-4(k^2-4)(2)=0\rightarrow 4k^2-8k^2+32=0\rightarrow 32=4k^2\rightarrow 8=k^2\rightarrow k=\pm 2\sqrt{2}$.

15. The equation for directrix is $4py=x^2$ where $p$ is the distance between the vertex and the directrix. Since $4p=\dfrac{1}{2}$, then $p=\dfrac{1}{8}$. Since the graph is concave up, the directrix must be $\dfrac{1}{8}$ below the vertex, so $y=-\dfrac{1}{8}$.

# Minitest 4

**Minitest 4** includes

- Trigonometric Identities
- Probabilities
- Space Geometry
- Polynomial Functions and Equations
- Even and Odd Functions
- Logarithm
- Radical Functions and Equations
- Conic Sections
- Inequalities
- Inverse Functions
- Linear Functions and Equations

# MATHEMATICS LEVEL 2 MINITEST 4

REFERENCE INFORMATION

THE FOLLOWING INFORMATION IS FOR YOUR REFERERENCE IN ANSWERING SOME OF THE QUESTIONS IN THE TEST.

Volume of a right circular cone with radius $r$ and height $h$ : $V = \frac{1}{3}\pi r^2 h$

Volume of a sphere with radius $r$ : $V = \frac{4}{3}\pi r^3$

Volume of a pyramid with base area $B$ and height $h$ : $V = \frac{1}{3}Bh$

Surface Area of a sphere with radius $r$ : $S = 4\pi r^2$

DO NOT DETACH FROM THE BOOK.

## MATHEMATICS LEVEL 2 MINITEST 4

For each of the following problems, decide which is the BEST of the choices given. If the exact numerical value is not one of the choices, select the choice that best approximates this value. Then fill in the corresponding circle on the answer sheet.

Notes: (1) A scientific or graphing calculator will be necessary for answering some (but not all) of the questions in this test. For each question you will have to decide whether or not you should use a calculator.

(2) For some questions in this test you may have to decide whether your calculator should be in the radian mode or the degree mode.

(3) Figures that accompany problems in this test are intended to provide information useful in solving the problems. They are drawn as accurately as possible EXCEPT when it is stated in a specific problem that its figure is not drawn to scale. All figures lie in a plane unless otherwise indicated.

(4) Unless otherwise specified, the domain of any function $f$ is assumed to be the set of all real numbers $x$ for which $f(x)$ is a real number. The range of $f$ is assumed to be the set of all real numbers $f(x)$, where $x$ is in the domain of $f$.

(5) Reference information that may be useful in answering the questions in this test can be found on the page preceding Question 1.

USE THIS SPACE FOR SCRATCH WORK.

1. $\sin(\theta - \frac{\pi}{2}) =$

(A) $\sin\theta$
(B) $-\sin\theta$
(C) $\cos\theta$
(D) $-\cos\theta$
(E) $\sin\theta\cos\theta$

**MATHEMATICS LEVEL 2 Test - *Continued***

USE THIS SPACE FOR SCRATCH WORK.

2. If the probability that a light bulb is defective is 0.1, what is the probability that a package of 10 light bulbs has exactly two defective bulbs?

(A) 0.01
(B) 0.10
(C) 0.19
(D) 0.25
(E) 0.33

3. What is the distance from the plane $3x-4y-5z+10=0$ to the point $(0,0,0)$?

(A) $\sqrt{2}$
(B) 2
(C) $2\sqrt{2}$
(D) 4
(E) $4\sqrt{2}$

4. If $f(x)=(x-1)(x^2+x+1)$, which of the following statements are true?

I. The function $f$ is increasing for $x \geq 1$.

II. The function $f(x)=0$ has three real solutions.

III. The domain of the function $f(x)$ is all real numbers.

(A) I only
(B) II only
(C) I and III only
(D) II and III only
(E) I, II, and III

**MATHEMATICS LEVEL 2 Test - *Continued***

USE THIS SPACE FOR SCRATCH WORK.

5. Which of the following is an even function?

(A) $f(x) = \sin(x)$
(B) $f(x) = \tan(x)$
(C) $f(x) = e^{2x}$
(D) $f(x) = 2x^2 - 3$
(E) $f(x) = \log(x)$

6. If $y = \log_5(x^2 - 6x + 14)$, what is the minimum value of the logarithmic function?

(A) $-2$ (B) $-1$ (C) 1 (D) 2 (E) 5

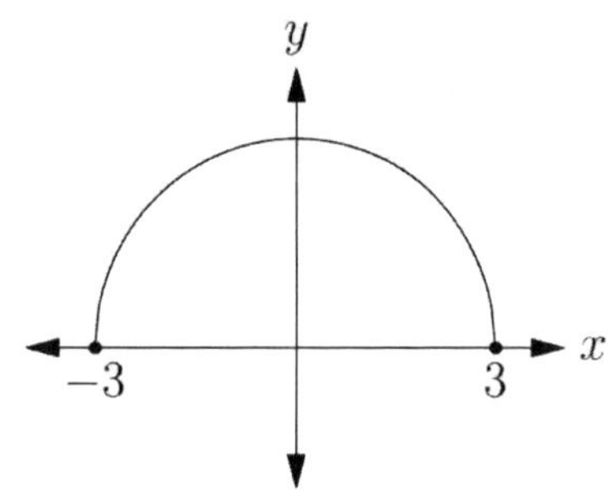


Figure of $y = \sqrt{9 - x^2}$

7. If $f(x) = \sqrt{9 - x^2}$ and $x \leq 0$, what is the inverse of $f(x)$?

(A) $f^{-1}(x) = \sqrt{x^2 - 3}$
(B) $f^{-1}(x) = \sqrt{9 - x^2}$ and $x \leq 0$
(C) $f^{-1}(x) = -\sqrt{9 - x^2}$ and $x \geq 0$
(D) $f^{-1}(x) = -\sqrt{9 - x^2}$ and $x \leq 0$
(E) $f^{-1}(x) = \sqrt{x^2 - 9}$ and $x \leq 0$

**MATHEMATICS LEVEL 2 Test - *Continued***

USE THIS SPACE FOR SCRATCH WORK.

8. What is the equation of line $l$ that is tangent to the circle $x^2+y^2=1$ and passes through the point $(0,2)$?

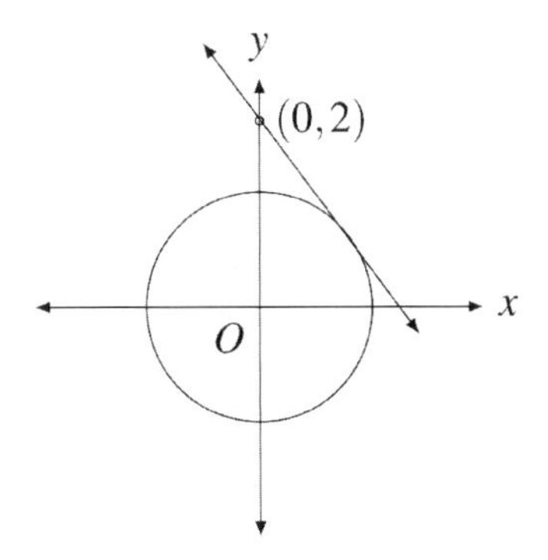


(A) $y=-x+2$
(B) $y=-\sqrt{2}x+2$
(C) $y=-\sqrt{3}x+2$
(D) $y=-2x+2$
(E) $y=-3x+2$

9. If $\dfrac{ab}{a-b}>0$, then which of the following must be true?

I. $0<b<a$

II. $b<a<0$

III. $a<b<0$

(A) I only
(B) I and II only
(C) II and III only
(D) I and III only
(E) I, II, and III

**MATHEMATICS LEVEL 2 Test - *Continued***

USE THIS SPACE FOR SCRATCH WORK.

10. If $f(x) = e^{2x}$ and $g(x) = \ln(x^2+1)$, then $(f \circ g)(x) =$

(A) $x^2+1$
(B) $x^3+x$
(C) $2x(x^2+1)$
(D) $x^4+2x^2+1$
(E) $2x\ln(x^2+1)$

11. A committee of 5 is to be chosen from 8 men and 5 women. What is the probability that the committee consists of 2 men and 3 women?

(A) 0.185
(B) 0.218
(C) 0.302
(D) 0.387
(E) 0.425

12. If $\log_2(\log_3(\log_2(x))) = 1$, what is the value of $x$?

(A) 126
(B) 256
(C) 512
(D) 1024
(E) 2048

## MATHEMATICS LEVEL 2 Test - *Continued*

USE THIS SPACE FOR SCRATCH WORK.

13. In a box there are 4 red marbles and 5 white marbles. If marbles are drawn one at a time and replaced after each drawing, what is the probability of drawing exactly 2 red marbles when 3 marbles are drawn?

(A) 0.329
(B) 0.235
(C) 0.198
(D) 0.110
(E) 0.102

14. $A, B$, and $C$, the vertices for the squares, are collinear. What is the value of $k$?

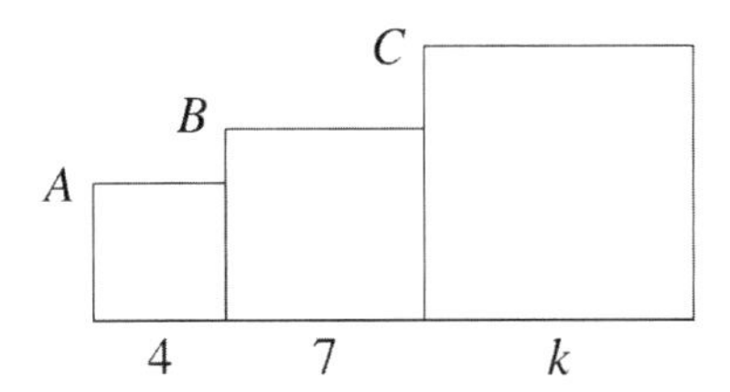


(A) 8.45
(B) 10.38
(C) 12.25
(D) 13.12
(E) 13.74

**MATHEMATICS LEVEL 2 Test - *Continued***

USE THIS SPACE FOR SCRATCH WORK.

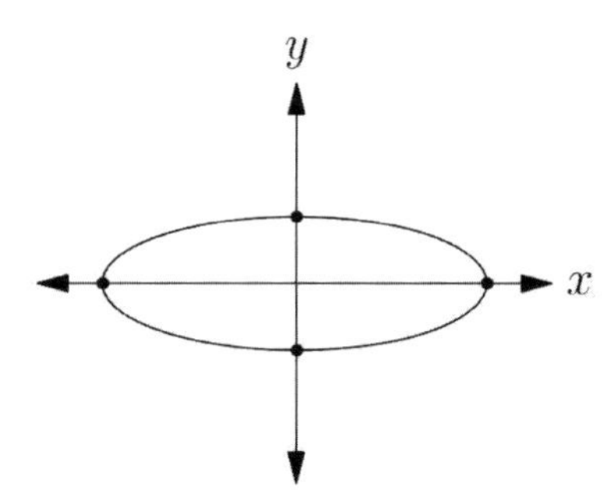


Figure of $5x^2 + 18y^2 - 90 = 0$

15. What is the length of the major axis of the ellipse whose equation is $5x^2 + 18y^2 - 90 = 0$?

(A) 6.25
(B) 7.25
(C) 8.49
(D) 9.34
(E) 10.25

## Answerkey to Minitest 4

1. (D)

2. (C)

3. (A)

4. (C)

5. (D)

6. (C)

7. (C)

8. (C)

9. (C)

10. (D)

11. (B)

12. (C)

13. (A)

14. (C)

15. (C)

## Solution

1. $\sin(\theta - \frac{\pi}{2}) = \sin(-(\frac{\pi}{2} - \theta)) = -\sin(\frac{\pi}{2} - \theta) = \cos(\theta)$.

2. $\binom{10}{2}\left(\frac{1}{10}\right)^2\left(\frac{9}{10}\right)^8 \approx 0.19$.

3. The distance between $(0,0,0)$ and $3x - 4y - 5z + 10 = 0$ is $\frac{|3(0) - 4(0) - 5(0) + 10|}{\sqrt{3^2 + (-4)^2 + (-5)^2}} = \sqrt{2}$.

4. $f(x) = (x-1)(x^2 + x + 1) = x^3 - 1$ is increasing for all $x$ with one $x$-intercept. Since this is a polynomial function, the domain is the set of all real numbers. Hence, I and III are true.

5. $y = 2x^2 - 3 = 2x^2 - 3x^0$ is even function because its exponents are all even.

6.
$\log_5(x^2 - 6x + 14) = \log_5(x^2 - 6x + 9 - 9 + 14) = \log_5(x^2 - 6x + 9 + 5) = \log_5((x^2 - 6x + 9) + 5) \geq \log_5(5) = 1$, so $y \geq 1$.

7. Reflecting the quarter circle in the second quadrant about the line $y = x$ results in the quarter circle in the fourth quadrant. Hence, $f^{-1}(x) = -\sqrt{9 - x^2}$ where $x \geq 0$.

8. If we substitute $y = -\sqrt{3}x + 2$ into $x^2 + y^2 = 1$, then
$x^2 + (-\sqrt{3}x + 2)^2 = 1 \rightarrow x^2 + 3x^2 - 4\sqrt{3}x + 4 = 1 \rightarrow 4x^2 - 4\sqrt{3}x + 3 = 0$. The discriminant is $(-4\sqrt{3})^2 - 4(3)(4) = 0$, meaning that the line is tangent to the curve.

9. If $\frac{ab}{a-b} < 0$, then $ab < 0, a - b < 0$ or $ab > 0, a - b > 0$. Then, the only possible cases are $a < 0 < b$ or $0 < b < a$ or $b < a < 0$. Hence, I and II are true.

10. $e^{2\ln(x^2+1)} = e^{\ln((x^2+1)^2)} = e^{\ln(x^4+2x^2+1)} = x^4 + 2x^2 + 1$.

11. $\frac{8}{13} \times \frac{7}{12} \times \frac{5}{11} \times \frac{4}{10} \times \frac{3}{9} \times \frac{5!}{2!3!} \approx 0.218$.

12. $\log_2(\log_3(\log_2(x))) = 1 \rightarrow \log_3(\log_2(x)) = 2 \rightarrow \log_2(x) = 9 \rightarrow x = 2^9 = 512$.

13. $\frac{4}{9} \times \frac{4}{9} \times \frac{5}{9} \times \frac{3!}{2!1!} \approx 0.329$.

14. Since $A$, $B$, and $C$ are collinear, the slope between $A$ and $B$ equals that between $B$ and $C$, so $\frac{3}{4} = \frac{k-7}{7}$, so $k = 12.25$.

15. $5x^2 + 18y^2 = 90 \rightarrow \frac{x^2}{18} + \frac{y^2}{5} = 1$, so the length of major axis is $2\sqrt{18} = 6\sqrt{2} \approx 8.49$.

## Point-Line Distance

$$\text{Distance} = \frac{|A(m)+B(n)+C|}{\sqrt{A^2+B^2}}$$

where $Ax+By+C=0$ is the line and $(m,n)$ is the point.

In order to deduce this formula, we must use vector knowledge about the normal line, which is bit beyond the scope of the test. Hence, let's have a look at application of this formula so that we may use it during the test.

If we have to compute the distance between the origin and the line $y=-4x+7$, then we switch $y=-4x+7$ into $y+4x-7=0$, so the distance between the origin and the line can be computed by

$$\begin{aligned}\text{Distance} &= \frac{|(0)+4(0)-7|}{\sqrt{1^2+4^2}} \\ &= \frac{7}{\sqrt{17}} \\ &= \frac{7\sqrt{17}}{17}\end{aligned}$$

# Minitest 5

**Minitest 5** includes

- Exponential Functions and Equations
- Probability
- Statistics
- Linear Functions and Equations
- Limits
- Quadratic Functions and Equations
- Radical Functions and Equations
- Logarithm
- System of Equations
- Complex Numbers
- Conic Sections

# MATHEMATICS LEVEL 2 MINITEST 5

REFERENCE INFORMATION

THE FOLLOWING INFORMATION IS FOR YOUR REFERERENCE IN ANSWERING SOME OF THE QUESTIONS IN THE TEST.

Volume of a right circular cone with radius $r$ and height $h$ : $V = \frac{1}{3}\pi r^2 h$

Volume of a sphere with radius $r$ : $V = \frac{4}{3}\pi r^3$

Volume of a pyramid with base area $B$ and height $h$ : $V = \frac{1}{3}Bh$

Surface Area of a sphere with radius $r$ : $S = 4\pi r^2$

DO NOT DETACH FROM THE BOOK.

## MATHEMATICS LEVEL 2 MINITEST 5

For each of the following problems, decide which is the BEST of the choices given. If the exact numerical value is not one of the choices, select the choice that best approximates this value. Then fill in the corresponding circle on the answer sheet.

Notes: (1) A scientific or graphing calculator will be necessary for answering some (but not all) of the questions in this test. For each question you will have to decide whether or not you should use a calculator.

(2) For some questions in this test you may have to decide whether your calculator should be in the radian mode or the degree mode.

(3) Figures that accompany problems in this test are intended to provide information useful in solving the problems. They are drawn as accurately as possible EXCEPT when it is stated in a specific problem that its figure is not drawn to scale. All figures lie in a plane unless otherwise indicated.

(4) Unless otherwise specified, the domain of any function $f$ is assumed to be the set of all real numbers $x$ for which $f(x)$ is a real number. The range of $f$ is assumed to be the set of all real numbers $f(x)$, where $x$ is in the domain of $f$.

(5) Reference information that may be useful in answering the questions in this test can be found on the page preceding Question 1.

USE THIS SPACE FOR SCRATCH WORK.

1. The half-life of a radioactive substance called "Niners" is 9 years. If 40 grams of the substance exist initially, how much will remain after 23.5 years?

(A) 0.077 grams
(B) 244.30 grams
(C) 6.11 grams
(D) 2.49 grams
(E) 6.55 grams

## MATHEMATICS LEVEL 2 Test - *Continued*

USE THIS SPACE FOR SCRATCH WORK.

2. In a class of 25 students, 80% are passing the class with a grade of $C$ or better. If two students are randomly selected from the class, what is the probability that neither student is passing with a grade of $C$ or better?

(A) 0.03
(B) 0.20
(C) 0.08
(D) 0.63
(E) 0.64

3. Given a set of ten scores of mock test results for SAT Math Level II, if each score in a set of scores is decreased by 2, which of the following MUST be true?

I. The mean is decreased by 2.

II. The mean is unchanged.

III. The standard deviation is decreased by 2.

(A) I only
(B) III only
(C) I and III only
(D) II and IV only
(E) I and IV only

4. A line satisfies parametric equations $x(t) = 8 - t$ and $y(t) = 10 + 2t$ where $t$ is a parameter. What is the slope of the line?

(A) $-1$ (B) $-2$ (C) 3 (D) 2 (E) $\frac{1}{2}$

**MATHEMATICS LEVEL 2 Test - *Continued***

USE THIS SPACE FOR SCRATCH WORK.

5. $\lim_{x \to 3} \frac{3x^2 - 7x - 6}{x^2 - 9} =$

(A) 0
(B) 1.6
(C) 1.8
(D) 2.4
(E) There is no limit.

6. The operation $\star$ is defined for all real numbers $a$ and $b$ by the equation $a \star b = a^{-b} - 3b$. If $n \star (-2) = 70$, which of the following could equal $n$?

(A) 7
(B) 9
(C) −8
(D) −9
(E) −7

7. $2,500 is invested at a rate of 4.5% compounded monthly. The value of the investment in $t$ years can be modeled by the equation

$$A(t) = 2,500(1 + \frac{0.045}{12})^{12t}$$

How long will it take for the investment to double?

(A) 10.2 years
(B) 15.4 years
(C) 18.8 years
(D) 25 years
(E) 185.2 years

**MATHEMATICS LEVEL 2 Test - *Continued***

USE THIS SPACE FOR SCRATCH WORK.

8. What is the range of $f(x) = \sqrt{4 - x^2}$?

(A) $y \geq 0$
(B) $y \geq 2$
(C) $-2 \leq y \leq 2$
(D) $0 \leq y \leq 2$
(E) $y \leq 2$

9. If $f(x) = \dfrac{4x^2 - 9}{2x + 3}$, what value does the function approach as $x$ approaches to $-\dfrac{3}{2}$?

(A) 0.02
(B) 0
(C) $-4.33$
(D) $-6$
(E) $-9$

10. Five blue marbles and six red marbles are held in a single container. Marbles are randomly selected one at a time and not returned to the container. If three marbles are selected at a time, what is the probability that at least two red marbles will be chosen?

(A) $\dfrac{5}{33}$
(B) $\dfrac{5}{11}$
(C) $\dfrac{6}{11}$
(D) $\dfrac{19}{33}$
(E) $\dfrac{2}{3}$

**MATHEMATICS LEVEL 2 Test - *Continued***

USE THIS SPACE FOR SCRATCH WORK.

11. If $x > 0$ and $y > 1$, then $\log_{x^2} y =$

I. $\log_x y^2$

II. $\log_x \sqrt{y}$

III. $\log_x \frac{y}{2}$

(A) I only
(B) II only
(C) III only
(D) I and II only
(E) II and III only

12. The system of equations given by

$$\begin{cases} 2x + 3y = 7 \\ 10x + cy = 3 \end{cases}$$

has solutions for all values of $c$ EXCEPT

(A) $-15$
(B) $-3$
(C) 3
(D) 10
(E) 15

**MATHEMATICS LEVEL 2 Test - *Continued***

USE THIS SPACE FOR SCRATCH WORK.

13. If $W$ is a complex number shown in the figure, which of the following could be $iW$?

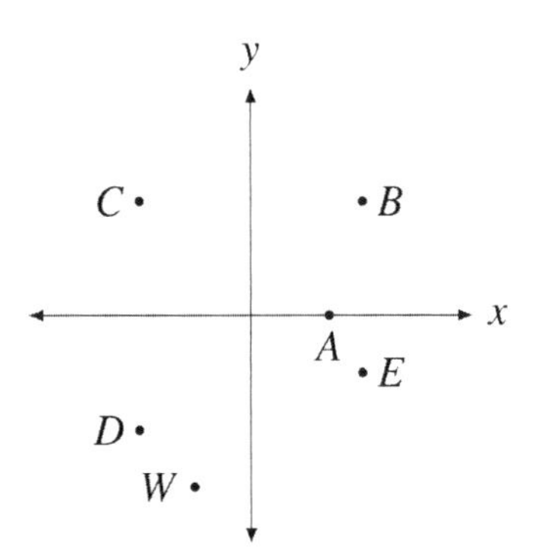


(A) $A$ (B) $B$ (C) $C$ (D) $D$ (E) $E$

14. In a room packed with people for social gathering, all people shook hands with each other exactly once. If twenty-one handshakes are exchanged, how many people were in the room?

(A) 5
(B) 6
(C) 7
(D) 8
(E) 9

15. What is the greatest possible number of points of intersection between a parabola and a circle?

(A) 2
(B) 3
(C) 4
(D) 6
(E) 8

## Answerkey to Minitest 5

1. (E)

2. (A)

3. (A)

4. (B)

5. (C)

6. (C)

7. (B)

8. (D)

9. (D)

10. (D)

11. (B)

12. (E)

13. (E)

14. (C)

15. (C)

## Solution

1. According to half-life equation, $40 \times (\frac{1}{2})^{\frac{23.5}{9}} \approx 6.55$.

2. $\frac{5}{25} \times \frac{4}{24} = \frac{1}{30} \approx 0.03$.

3. The mean changes but standard deviation does not change if each of the data is added or subtracted by a constant. Hence, I is true.

4. $\frac{\triangle y}{\triangle x} = \frac{2}{-1} = -2$.

5. $\lim_{x\to 3} \frac{(3x+2)(x-3)}{(x-3)(x+3)} = \lim_{x\to 3} \frac{3x+2}{x+3} = \frac{3(3)+2}{3+3} = \frac{11}{6} \approx 1.8$.

6. $n \star (-2) = 70 \rightarrow n^2 + 6 = 70 \rightarrow n^2 = 64 \rightarrow n = \pm 8$.

7.
$$2,500(1+\frac{0.045}{12})^{12t} = 5,000 \rightarrow (1+\frac{0.045}{12})^{12t} = 2 \rightarrow 12t = \frac{\log(2)}{\log(1+\frac{0.045}{12})} \rightarrow t = \frac{1}{12} \times \frac{\log(2)}{\log(1+\frac{0.045}{12})} \approx 15.4.$$

8. The graph of $y = \sqrt{4-x^2}$ is the upper semicircle of a circle centered at $(0,0)$ with the radius of 2. Hence, $0 \le y \le 2$.

9. $\lim_{x\to -\frac{3}{2}} \frac{4x^2-9}{2x+3} = \lim_{x\to -\frac{3}{2}} (2x-3) = 2(-\frac{3}{2}) - 3 = -6$.

10. There are two possible cases. The first case is when there are two red marbles selected, which is equal to $\frac{6}{11} \times \frac{5}{10} \times \frac{5}{9} \times \frac{3!}{2!1!}$. The second case is when there are three marbles selected, which is equal to $\frac{6}{11} \times \frac{5}{10} \times \frac{4}{9}$. The sum of the two cases is $\frac{19}{33}$.

11. $\log_{x^2} y = \frac{1}{2} \log_x y = \log_x(y^{1/2}) = \log_x(\sqrt{y})$. Hence, II is true.

12. The system of equation has a solution if $\frac{2}{10} \neq \frac{3}{c}$, so $2c \neq 30$. Hence, $c \neq 15$.

13. Suppose $W = -1-3i$, for instance. Then, $iW = (-1-3i)i = -i+3$, which is roughly (E). Or, you could use the fact that multiplying by $i$ to any complex number is $90°$ counterclockwise rotation about the origin.

14. $\binom{n}{2} = 21 \rightarrow \frac{n(n-1)}{2} = 21 \rightarrow n(n-1) = 42 \rightarrow n = 7$.

15. The maximum number of intersection points between a parabola and a circle is 4.

# Minitest 6

**Minitest 6** includes

- Combinations and Permutations
- Polynomial Functions and Equations
- Periodic Functions
- Linear Functions and Equations
- Trigonometric Ratio
- Rational Functions and Equations
- Space Geometry
- Piecewise Functions
- Logarithm
- Statistics

# MATHEMATICS LEVEL 2 MINITEST 6

REFERENCE INFORMATION

THE FOLLOWING INFORMATION IS FOR YOUR REFERERENCE IN ANSWERING SOME OF THE QUESTIONS IN THE TEST.

Volume of a right circular cone with radius $r$ and height $h$ : $V = \frac{1}{3}\pi r^2 h$

Volume of a sphere with radius $r$ : $V = \frac{4}{3}\pi r^3$

Volume of a pyramid with base area $B$ and height $h$ : $V = \frac{1}{3}Bh$

Surface Area of a sphere with radius $r$ : $S = 4\pi r^2$

DO NOT DETACH FROM THE BOOK.

## MATHEMATICS LEVEL 2 MINITEST 6

For each of the following problems, decide which is the BEST of the choices given. If the exact numerical value is not one of the choices, select the choice that best approximates this value. Then fill in the corresponding circle on the answer sheet.

Notes: (1) A scientific or graphing calculator will be necessary for answering some (but not all) of the questions in this test. For each question you will have to decide whether or not you should use a calculator.

(2) For some questions in this test you may have to decide whether your calculator should be in the radian mode or the degree mode.

(3) Figures that accompany problems in this test are intended to provide information useful in solving the problems. They are drawn as accurately as possible EXCEPT when it is stated in a specific problem that its figure is not drawn to scale. All figures lie in a plane unless otherwise indicated.

(4) Unless otherwise specified, the domain of any function $f$ is assumed to be the set of all real numbers $x$ for which $f(x)$ is a real number. The range of $f$ is assumed to be the set of all real numbers $f(x)$, where $x$ is in the domain of $f$.

(5) Reference information that may be useful in answering the questions in this test can be found on the page preceding Question 1.

1. Nine different books written by different authors are put on an old book shelf made with oak trees. As Bob moves out, he must package all the books into three identical-looking paper bags. What is the number of all possible cases for Bob to put 9 books in three bags, each with 4, 3, and 2 books?

(A) 180
(B) 320
(C) 640
(D) 840
(E) 1,260

USE THIS SPACE FOR SCRATCH WORK.

**MATHEMATICS LEVEL 2 Test - *Continued***

USE THIS SPACE FOR SCRATCH WORK.

2. Which of the following satisfies the statement "if $a < b$, then $f(a) < f(b)$"?

(A) $y = |x|$
(B) $y = \sin(x)$
(C) $y = -x^2$
(D) $y = 2^{-x}$
(E) $y = (x-3)^5$

3. If $f(x)$ is a periodic function with the period of 4, and $f(x) = -2$, $f(x+1) = 0$, $f(x+2) = 2$ and $f(x+3) = 1$, then find $f(x+99)$.

(A) $-2$ (B) 0 (C) 1 (D) 2 (E) 4

4. In a 120 mile trip, a car was driving at the speed of 40 mile per hour until certain time $t$. After such $t$ hours, if the car traveled rest of the trip at 70 mile per hour, find the equation that represents the rest of the trip time after $t$ hours.

(A) $\dfrac{4t-12}{7}$
(B) $\dfrac{12t}{7}$
(C) $\dfrac{4t}{7}$
(D) $\dfrac{12-4t}{7}$
(E) $\dfrac{12}{7}$

**MATHEMATICS LEVEL 2 Test - *Continued***

USE THIS SPACE FOR SCRATCH WORK.

5. If $\sin(\theta) = \frac{x}{3}$, where $0° < \theta < 90°$, and $0 < x < 3$, then $\cos(\theta) =$

(A) $\frac{\sqrt{9-x^2}}{3}$

(B) $\frac{\sqrt{x^2-9}}{x}$

(C) $\frac{\sqrt{9-x^2}}{x}$

(D) $\frac{\sqrt{3-x^2}}{3}$

(E) $\frac{\sqrt{3-x^2}}{x}$

6. Which of the following must be true about $f(x) = ax^5 + bx^4 + c$ where $a \neq 0, b \neq 0$ and $c \neq 0$?

I. There is at least one $x$-intercept.

II. The domain is $\mathbb{R}$.

III. $f(1) > 0$

(A) I only
(B) II only
(C) III only
(D) I and II only
(E) II and III only

**MATHEMATICS LEVEL 2 Test - *Continued***

USE THIS SPACE FOR SCRATCH WORK.

7. A number of friends were to pay $270 for the total hotel fee on a trip. However, at the last moment, four friends did not show up. Thus, the rest of the members each needs to pay $24 more for the hotel. How much is the final cost of the hotel fee for each person?

(A) $30
(B) $34
(C) $42
(D) $48
(E) $54

8. James and his friends decided to participate in a Triathlon. His speed on these things were 2 miles per hour in the $\frac{1}{2}$ mile walk and 6 miles per hour in the 3 mile run. His goal is to finish the race in 3 hours. What must be his average speed, in miles per hour, for the 18 mile bicycle ride?

(A) 2
(B) 6
(C) 8
(D) 10
(E) 12

9. If $\cos(2x) = k$ where $x$ is an acute angle, which of the following correctly represents $2\sin^2(x)$?

(A) $k$
(B) $1-k$
(C) $k+1$
(D) $\frac{1-k}{2}$
(E) $k^2-1$

**MATHEMATICS LEVEL 2 Test - *Continued***

USE THIS SPACE FOR SCRATCH WORK.

10. If the center of the circle is $(2,3)$, and it is tangent to the $y$-axis, what is the slope of the line $l$?

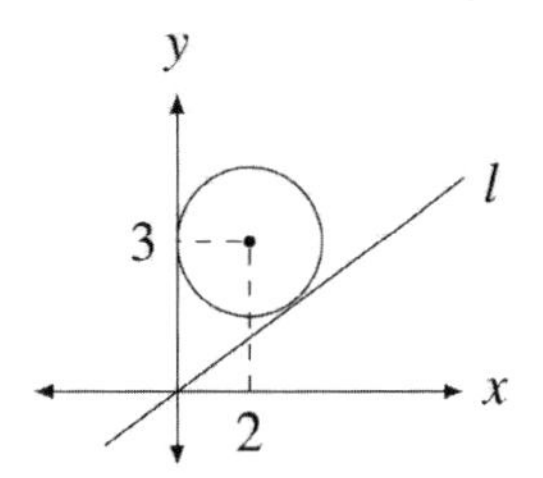


(A) $\frac{2}{3}$ (B) $\frac{3}{4}$ (C) $\frac{5}{12}$ (D) $\frac{3}{2}$ (E) $\frac{5}{6}$

11. A point $X$ in space is above the plane $Y$, where the point is at least 5 cm and at most 9 cm away from plane $Y$. Then, a circle is formed on the plane $Y$, whose area equals

(A) 75.24
(B) 120.80
(C) 134.35
(D) 175.93
(E) 203.47

12. What is the area of polygon formed by the points $(x, y)$ satisfying the inequality $2|x| + |y| \leq 2$?

(A) 2
(B) 3
(C) 4
(D) 8
(E) 10

**MATHEMATICS LEVEL 2 Test - *Continued***

USE THIS SPACE FOR SCRATCH WORK.

13. Which of the following could be the solution to $[n]^2 - 2[n] = 3$, where $[n]$ is the greatest integer function?

(A) $2 \leq n < 3$
(B) $0 \leq n < 1$
(C) $-1 < n < 0$
(D) $-1 \leq n < 0$
(E) $-1 < n \leq 0$

14. Which of the following is the solution set of the equation $\log_3(x-5) = \log_9(2x+5)$?

(A) $\{2\}$
(B) $\{5\}$
(C) $\{10\}$
(D) $\{2, 10\}$
(E) $\{2, 5, 10\}$

15. On your school math exam, the scores of ten students were 66, 81, 85, 97, 86, 58, 76, 73, 88, and 80. What is the standard deviation of the scores?

(A) 10.72
(B) 11.30
(C) 12.88
(D) 13.16
(E) 13.58

## Answerkey to Minitest 6

1. (E)
2. (E)
3. (C)
4. (D)
5. (A)
6. (D)
7. (E)
8. (C)
9. (B)
10. (C)
11. (D)
12. (C)
13. (D)
14. (C)
15. (A)

## Solution

1. $\binom{9}{4} \times \binom{5}{3} \times \binom{2}{2} = 1,260.$

2. If $a < b$, then $f(a) < f(b)$, where $f(x)$ is increasing function. Hence, the only increasing function in the answer choices is $y = (x-3)^5$.

3. $f(x+99) = f(x+95) = \cdots = f(x+3) = 1.$

4. $40t + 70\square = 120 \rightarrow 70\square = 120 - 40t \rightarrow \square = \dfrac{120-40t}{70} \rightarrow \square = \dfrac{12-4t}{7}.$

5. $\cos(\theta) = \sqrt{1-\sin^2(\theta)} = \sqrt{1-\dfrac{x^2}{9}} = \sqrt{\dfrac{9-x^2}{9}} = \dfrac{\sqrt{9-x^2}}{3}.$

6. Since the degree of $f(x)$ is 5, there should be at least one $x$-intercept. Also, the domain is the set of all real numbers because it is a polynomial function. However, $f(1) = a+b+c$ is not necessarily greater than 0. Hence, I and II are true.

7. Let there be $n$ number of friends. Then, $\dfrac{270}{n} + 24 = \dfrac{270}{n-4}$, so $n = 9$. Hence, at the moment of traveling, there are five friends paying for the hotel fee. Hence, $\dfrac{\$270}{5} = \$54$ is the final cost of the hotel fee for each person.

8. $\dfrac{1/2}{2} + \dfrac{3}{6} + \dfrac{18}{x} = 3$. Hence, $x = 8$ miles per hour for the 18 mile bicycle ride.

9. $\cos(2x) = 1 - 2\sin^2(x) = k$. Hence, $1 - k = 2\sin^2(x)$.

10.

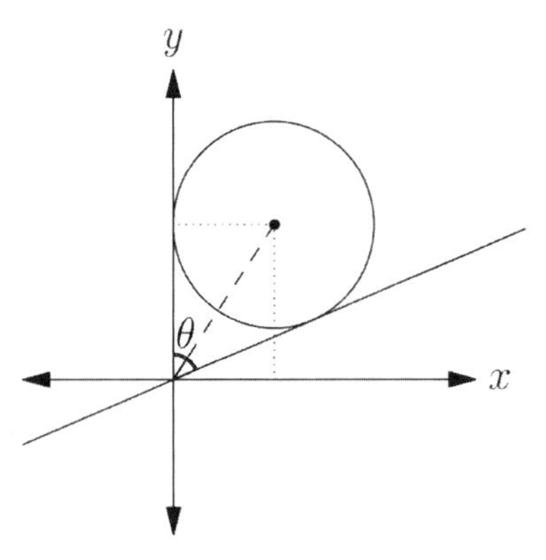


Since $\tan(\theta) = \dfrac{2}{3}$, we compute $\tan(90-2\theta) = \cot(2\theta) = \dfrac{1-\tan^2(\theta)}{2\tan(\theta)} = \dfrac{1-4/9}{4/3} = \dfrac{5}{12}$, which is the slope of the line tangent to the circle.

11. The radius of the circle on the plane is $\sqrt{9^2-5^2} = \sqrt{81-25} = \sqrt{56}$. Hence, the area of the circle on the plane is $56\pi \approx 175.93$.

12. The polygon is a rhombus whose diagonal has the length of 2 and 4. Hence, its area is $\dfrac{4\times 2}{2} = 4$.

13. $\lfloor n \rfloor^2 - 2\lfloor n \rfloor n - 3 = 0 \implies \lfloor n \rfloor = 3$ or $-1 \implies 3 \le n < 4$ or $-1 \le n < 0$.

14. $\log_3(x-5) = \log_9(2x+5) \implies \log_3(x-5) = \log_3\sqrt{2x+5} \implies x-5 = \sqrt{2x+5} \implies x = 10$.

15. The mean value of the test scores is $\frac{66+81+85+97+86+58+76+73+88+80}{10} = 79$. Hence, the standard deviation of the ten scores is given by

$$\sqrt{\frac{(66-79)^2+(81-79)^2+(85-79)^2+\cdots+(80-79)^2}{10}} = \sqrt{115} \approx 10.72$$

# CHECK ON LEARNING #11

## Sum of Tangents

$$\tan(A+B) = \frac{\tan(A)+\tan(B)}{1-\tan(A)\tan(B)}$$

Let's deduce the formula by using $\tan(x) = \frac{\sin(x)}{\cos(x)}$.

$$\begin{aligned}\tan(A+B) &= \frac{\sin(A+B)}{\cos(A+B)} \\ &= \frac{\sin(A)\cos(B)+\cos(A)\sin(B)}{\cos(A)\cos(B)-\sin(A)\sin(B)} \\ &= \frac{\sin(A)/\cos(A)+\sin(B)/\cos(B)}{1-\tan(A)\tan(B)} \\ &= \frac{\tan(A)+\tan(B)}{1-\tan(A)\tan(B)}\end{aligned}$$

This formula is useful when we have to deal with slopes of linear function. Remember that

$$\tan(\theta) = \frac{\triangle y}{\triangle x}$$

where $\theta$ is the angle formed by the line and the $x$-axis.

# Minitest 7

**Minitest 7** includes

- Inverse Functions
- Parametric Functions
- Polar Coordinates
- Even and Odd Functions
- Conic Sections
- Radical Functions and Equations
- Logarithm
- Quadratic Functions and Equations
- Trigonometric Functions
- Sequence and Series

# MATHEMATICS LEVEL 2 MINITEST 7

REFERENCE INFORMATION

THE FOLLOWING INFORMATION IS FOR YOUR REFERERENCE IN ANSWERING SOME OF THE QUESTIONS IN THE TEST.

Volume of a right circular cone with radius $r$ and height $h$ : $V = \frac{1}{3}\pi r^2 h$

Volume of a sphere with radius $r$ : $V = \frac{4}{3}\pi r^3$

Volume of a pyramid with base area $B$ and height $h$ : $V = \frac{1}{3}Bh$

Surface Area of a sphere with radius $r$ : $S = 4\pi r^2$

DO NOT DETACH FROM THE BOOK.

## MATHEMATICS LEVEL 2 MINITEST 7

For each of the following problems, decide which is the BEST of the choices given. If the exact numerical value is not one of the choices, select the choice that best approximates this value. Then fill in the corresponding circle on the answer sheet.

Notes: (1) A scientific or graphing calculator will be necessary for answering some (but not all) of the questions in this test. For each question you will have to decide whether or not you should use a calculator.

(2) For some questions in this test you may have to decide whether your calculator should be in the radian mode or the degree mode.

(3) Figures that accompany problems in this test are intended to provide information useful in solving the problems. They are drawn as accurately as possible EXCEPT when it is stated in a specific problem that its figure is not drawn to scale. All figures lie in a plane unless otherwise indicated.

(4) Unless otherwise specified, the domain of any function $f$ is assumed to be the set of all real numbers $x$ for which $f(x)$ is a real number. The range of $f$ is assumed to be the set of all real numbers $f(x)$, where $x$ is in the domain of $f$.

(5) Reference information that may be useful in answering the questions in this test can be found on the page preceding Question 1.

USE THIS SPACE FOR SCRATCH WORK.

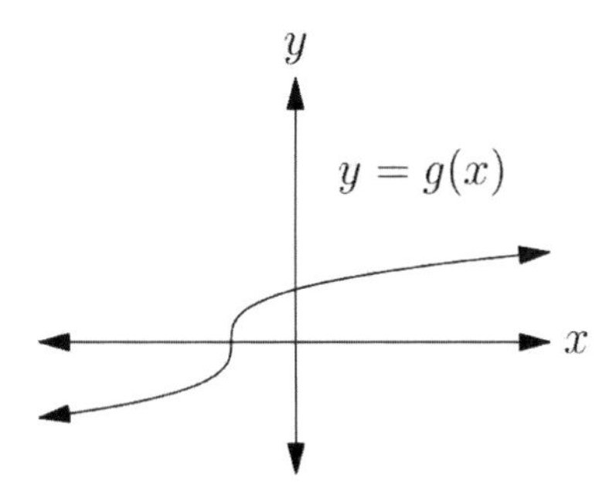


1. If $g(x) = \sqrt[3]{\dfrac{x+1}{2}}$, then $g^{-1}(1.5) =$

(A) 4.25
(B) 5.75
(C) 6.52
(D) 7.12
(E) 8.45

**MATHEMATICS LEVEL 2 Test - *Continued***

USE THIS SPACE FOR SCRATCH WORK.

2. What is the graph of the parametrically defined equations $x(\theta) = 4 + 2\cos\theta$ and $y(\theta) = \sin\theta - 1$?

(A) Parabola
(B) Circle
(C) Ellipse
(D) Hyperbola
(E) Two perpendicular lines

3. What is the polar form of the rectangular equation $x^2 + y^2 - 4x = 0$?

(A) $r = \sin\theta$
(B) $r^2 = 4\sin\theta$
(C) $r = 4\cos\theta$
(D) $r = 4\sin\theta$
(E) $r = 2\cos\theta$

4. If $f(x) = f(-x)$ for all real $x$ and a point (3, 5) is on the line, then which of the following points is also on the line?

(A) $(-3, 5)$
(B) $(3, -5)$
(C) $(-5, -3)$
(D) $(-3, -5)$
(E) $(5, 3)$

**MATHEMATICS LEVEL 2 Test - *Continued***

USE THIS SPACE FOR SCRATCH WORK.

5. If $y = f(x)$ is symmetric with respect to $x = 7$ and $(2,5)$ is on its graph, which of the following point must be on $y = f(x)$?

(A) $(-2,5)$
(B) $(2,9)$
(C) $(12,5)$
(D) $(7,5)$
(E) $(-5,5)$

6. If two circles are symmetric with respect to $y = x$, and one of the circle satisfies the equation $x^2 + y^2 - 2x - 4y + 1 = 0$, then which of the following is the equation of the other circle?

(A) $(x-2)^2 + (y-1)^2 = 4$
(B) $(x-1)^2 + (y-2)^2 = 2$
(C) $(x-1)^2 + (y-2)^2 = 4$
(D) $(x-2)^2 + (y-2)^2 = 4$
(E) $(x+1)^2 + (y+2)^2 = 4$

7. If $x < 3$, then $\sqrt{(x-10)^2} =$

(A) $10 - x$
(B) $10 + x$
(C) $x - 10$
(D) $-x - 10$
(E) $(x-10)^2$

**MATHEMATICS LEVEL 2 Test - *Continued***

USE THIS SPACE FOR SCRATCH WORK.

8. What is the range of the function $f(x) = -\sqrt{3x-9}+4$?

(A) $y \geq 3$
(B) $y \leq 3$
(C) $y \geq 4$
(D) $y \leq 4$
(E) $y \leq -4$

9. If $\log_{\sqrt{3}} x = 10$, then $\log_3 x^3 =$

(A) 10
(B) 15
(C) 30
(D) 45
(E) 60

10. If $f(x) = \log(x+1) + \log(x-1)$, then $f^{-1}(x) =$

(A) $10^{x^2-1}$
(B) $x^2 - 10$
(C) $\sqrt{10^x+1}$
(D) $-\sqrt{10^x+1}$
(E) $\sqrt{10^x-1}$

**MATHEMATICS LEVEL 2 Test - *Continued***

USE THIS SPACE FOR SCRATCH WORK.

11. If $f(x) = \sqrt[3]{2x+3}$, then $f^{-1}(3) =$

(A) 3
(B) 5
(C) 8
(D) 10
(E) 12

12. If the difference of integer roots of $x^2 + 2mx = 7$ is 8, then what is the positive integer value of $m$?

(A) 0
(B) 1
(C) 2
(D) 3
(E) 4

13. $\log_3(ab) = 10$ and $\log_3\left(\frac{a}{b}\right) = 2$, and $b > 0$, then what is the value of $a$?

(A) 9
(B) 81
(C) 243
(D) 729
(E) 2187

**MATHEMATICS LEVEL 2 Test - *Continued***

USE THIS SPACE FOR SCRATCH WORK.

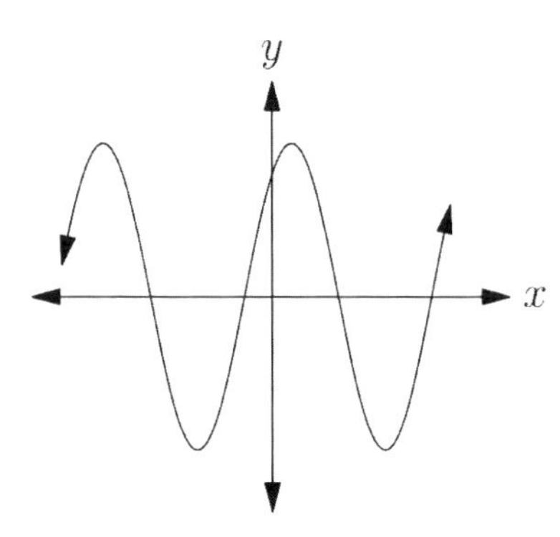


Figure of $y = 3\sin(x) + 4\cos(x)$

14. If $\sin(A+B) = \sin(A)\cos(B) + \cos(A)\sin(B)$ for any real $A$ and $B$, what is the amplitude of $y = 3\sin(x) + 4\cos(x)$?

(A) 3
(B) 4
(C) 5
(D) 6
(E) 7

15. If the sum of the first $n$ terms of a series is $S_n = n^2 + 4n$, then what is the 10th term?

(A) 23
(B) 40
(C) 85
(D) 125
(E) 140

## Answerkey to Minitest 7

1. (B)

2. (C)

3. (C)

4. (A)

5. (C)

6. (A)

7. (A)

8. (D)

9. (B)

10. (C)

11. (E)

12. (D)

13. (D)

14. (C)

15. (A)

## Solution

1. Let $g^{-1}(1.5) = x$. Then,
$1.5 = \sqrt[3]{\frac{x+1}{2}} \implies (1.5)^3 = \frac{x+1}{2} \implies (1.5)^3 \times 2 = x+1 \implies (1.5)^3 \times 2 - 1 = x \implies x = 5.75.$

2. Since $x-4 = 2\cos(\theta), y+1 = \sin(\theta)$, then $\frac{(x-4)^2}{4} + \frac{(y+1)^2}{1} = \cos^2(\theta) + \sin^2(\theta) = 1$. Hence, the graph is ellipse.

3. Since $x^2 + y^2 = r^2$ and $x = r\cos(\theta)$, then $r^2 = 4r\cos(\theta) \implies r = 4\cos(\theta)$.

4. Since $f(3) = 5$, then $f(-3) = 5$ by $f(x) = f(-x)$.

5. $(2,5)$ is reflected about the line $x = 7$ to reach $(12,5)$.

6. $x^2 + y^2 - 2x - 4y + 1 = 0 \implies (x^2 - 2x + 1) + (y^2 - 4y + 4) = -1 + 1 + 4 = 4 \implies (x-1)^2 + (y-2)^2 = 4.$

7. $\sqrt{(x-10)^2} = |x - 10| \implies -(x-10)$ for $x < 10 \implies |x-10| = 10 - x$ for $x < 10$.

8. For all $x$ in the domain, $\sqrt{3x-9} \geq 0 \implies -\sqrt{3x-9} \leq 0 \implies -\sqrt{3x-9} + 4 \leq 4$.

9. $\log_{\sqrt{3}}(x) = 10 \implies (\sqrt{3})^{10} = 3^5 = x \implies \log_3(3^5)^3 \implies \log_3(3^{15}) = 15.$

10. $f(x) = \log(x+1) + \log(x-1) = \log(x^2-1) \implies f(f^{-1}(x)) = \log((f^{-1}(x))^2 - 1) = x \implies 10^x + 1 = (f^{-1}(x))^2 \implies \sqrt{10^x+1} = f^{-1}(x)$.

11. Let $f^{-1}(3) = x$. Then,
$3 = f(x) \implies 3 = \sqrt[3]{2x+3} \implies 3^3 = 2x+3 \implies 3^3 - 3 = 2x \implies 24 = 2x \implies x = 12.$

12. $x^2 + 2mx - 7 = 0$, so the possible roots are $(-1,7)$ or $(1,-7)$. Hence,
$2m = 6$ or $-6 \implies m = 3$ or $-3$.

13. $\log_3(a) + \log_3(b) = 10$ and $\log_3(a) - \log_3(b) = 2$. Therefore, $2\log_3(a) = 12$, so $\log_3(a) = 6$. Hence, $a = 3^6 = 27^2 = 729$.

14. $y = 3\sin(x) + 4\cos(x) = 5\left(\frac{3}{5}\sin(x) + \frac{4}{5}\cos(x)\right) = 5\sin(x + 53.13°)$. Hence, the amplitude is 5.

15. Since $S_{10} = a_1 + a_2 + a_3 + \cdots + a_{10}$ and $S_9 = a_1 + a_2 + \cdots + a_9$, so
$a_{10} = S_{10} - S_9 = (10^2 + 4(10)) - (9^2 + 4(9)) = 140 - 117 = 23.$

# Minitest 8

**Minitest 8** includes

- Trigonometric Functions
- Trigonometric Identities
- Sequence and Series
- Radical Functions and Equations
- Combinations and Permutations
- Vectors
- Conic Sections
- Complex Numbers
- Rational Functions and Equations
- Piecewise Functions and Equations

# MATHEMATICS LEVEL 2 MINITEST 8

REFERENCE INFORMATION

THE FOLLOWING INFORMATION IS FOR YOUR REFERERENCE IN ANSWERING SOME OF THE QUESTIONS IN THE TEST.

Volume of a right circular cone with radius $r$ and height $h$ : $V = \frac{1}{3}\pi r^2 h$

Volume of a sphere with radius $r$ : $V = \frac{4}{3}\pi r^3$

Volume of a pyramid with base area $B$ and height $h$ : $V = \frac{1}{3}Bh$

Surface Area of a sphere with radius $r$ : $S = 4\pi r^2$

DO NOT DETACH FROM THE BOOK.

## MATHEMATICS LEVEL 2 MINITEST 8

For each of the following problems, decide which is the BEST of the choices given. If the exact numerical value is not one of the choices, select the choice that best approximates this value. Then fill in the corresponding circle on the answer sheet.

Notes: (1) A scientific or graphing calculator will be necessary for answering some (but not all) of the questions in this test. For each question you will have to decide whether or not you should use a calculator.

(2) For some questions in this test you may have to decide whether your calculator should be in the radian mode or the degree mode.

(3) Figures that accompany problems in this test are intended to provide information useful in solving the problems. They are drawn as accurately as possible EXCEPT when it is stated in a specific problem that its figure is not drawn to scale. All figures lie in a plane unless otherwise indicated.

(4) Unless otherwise specified, the domain of any function $f$ is assumed to be the set of all real numbers $x$ for which $f(x)$ is a real number. The range of $f$ is assumed to be the set of all real numbers $f(x)$, where $x$ is in the domain of $f$.

(5) Reference information that may be useful in answering the questions in this test can be found on the page preceding Question 1.

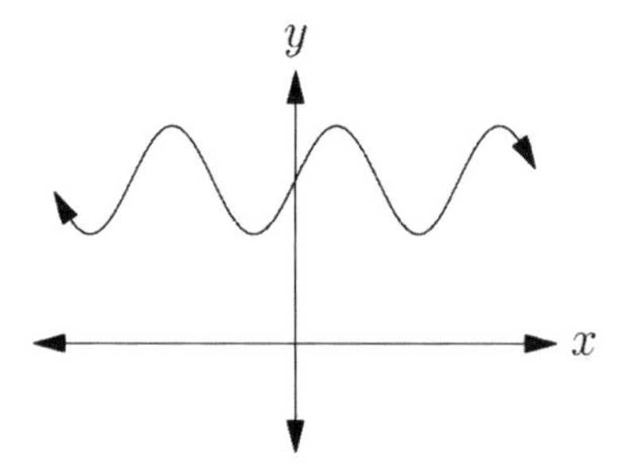


USE THIS SPACE FOR SCRATCH WORK.

1. Which of the following is true for the graph of $y = \sin(2x) + 3$?

(A) Symmetric with respect to the $y$-axis.
(B) Symmetric with respect to the $x$ axis.
(C) Symmetric with respect to the origin.
(D) Symmetric with respect to the point $(3,0)$
(E) Symmetric with respect to the point $(0,3)$

## MATHEMATICS LEVEL 2 Test - *Continued*

USE THIS SPACE FOR SCRATCH WORK.

2. If $\theta$ is a positive acute angle, and $\sin(2\theta) = \frac{1}{2}$, then $(\sin\theta + \cos\theta)^2 =$

(A) 1
(B) 1.5
(C) 2.2
(D) 2.5
(E) 2.8

3. Find the sum of the first 30 terms of the recursive sequence defined as $a_1 = 3$ and $a_n = a_{n-1} + 5$.

(A) 148
(B) 1680
(C) 2265
(D) 2340
(E) 3120

4. What is the period of the function defined by $f(x) = 5 - 2\cos^2\left(\frac{\pi x}{3}\right)$?

(A) 1
(B) 2
(C) 3
(D) 6
(E) 8

**MATHEMATICS LEVEL 2 Test -** ***Continued***

USE THIS SPACE FOR SCRATCH WORK.

5. What is the domain of the function $f(x) = \sqrt{8-2x-x^2}$?

(A) $(-\infty, -4]$
(B) $(-\infty, 5]$
(C) $[-4, 2]$
(D) $[-1, 4]$
(E) All real numbers

6. Six students are to be seated in a row of 6 chairs. If three of these students must be seated together, how many ways could this be accomplished?

(A) 24 (B) 48 (C) 120 (D) 144 (E) 210

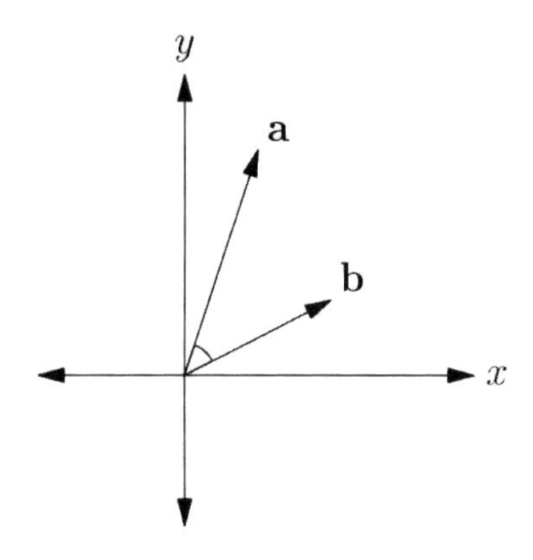


7. What is the measure of the angle between two vectors $\mathbf{a} = (1, 3)$ and $\mathbf{b} = (2, 1)$?

(A) 22.5°
(B) 30°
(C) 45°
(D) 60°
(E) 75°

**MATHEMATICS LEVEL 2 Test - *Continued***

USE THIS SPACE FOR SCRATCH WORK.

8. If a hyperbola has the equation $y^2 - 25x^2 = 25$, which of the following could be its asymptote equation?

(A) $y = \frac{1}{5}x$

(B) $y = \frac{2}{5}x$

(C) $y = 5x$

(D) $y = -\frac{1}{5}x$

(E) $y = -\frac{2}{5}x$

9. What is the product of $3 - 4i$ and its conjugate?

(A) $-7$
(B) 7
(C) $-25$
(D) 25
(E) 6

10. What is the domain of $f(x) = \sqrt{\frac{x}{10-x}}$?

(A) $x \geq 0$
(B) $0 < x < 10$
(C) $0 < x \leq 10$
(D) $0 \leq x \leq 10$
(E) $0 \leq x < 10$

**MATHEMATICS LEVEL 2 Test - *Continued***

USE THIS SPACE FOR SCRATCH WORK.

11. Which of the following could be the asymptote of $f(x) = \frac{2x^2 - 3x + 1}{x+1}$?

(A) $x = 1$
(B) $y = 2$
(C) $y = 2x$
(D) $y = 2x - 5$
(E) $y = 2x - 10$

12. Which of the following is not a function?

(A) $y = e^{-x}$
(B) $y = x^5$
(C) $xy = -9$
(D) $y = \sqrt{10 - x^2}$
(E) $y^2 = 4x$

13. $\frac{\sin 2\theta}{1 - \cos 2\theta} =$

(A) $\sin\theta$
(B) $\cos\theta$
(C) $\tan\theta$
(D) $\cot\theta$
(E) $\sec\theta$

**MATHEMATICS LEVEL 2 Test - *Continued***

USE THIS SPACE FOR SCRATCH WORK.

14. The minimum value of the function defined as

$$f(x) = \begin{cases} |x| & \text{if } |x| \leq 2 \\ 2 & \text{if } |x| > 2 \end{cases}$$

is equal to

(A) 4
(B) 2
(C) 0
(D) $-1$
(E) $-2$

15. The range of the function

$$f(x) = \frac{2x^2 - 5x - 1}{x + 3}$$

is the set of $y$ values whose interval expression equals to

(A) $(0, \infty)$
(B) $(-\infty, \infty)$
(C) $(-\infty, 2] \cup [10, +\infty)$
(D) $(-\infty, -33] \cup [-1, +\infty)$
(E) $(-\infty, 2]$

## Answerkey to Minitest 8

1. (E)

2. (B)

3. (C)

4. (C)

5. (C)

6. (D)

7. (C)

8. (C)

9. (D)

10. (E)

11. (D)

12. (E)

13. (D)

14. (C)

15. (D)

## Solution

1. Since the graph of $y=\sin(2x)$ is symmetric about the origin, the graph of $y=\sin(2x)+3$ is symmetric about the point $(0,3)$.

2. $(\sin\theta+\cos\theta)^2=\sin^2(\theta)+2\sin(\theta)\cos(\theta)+\cos^2(\theta)=1+2\sin(\theta)\cos(\theta)=1+\sin(2\theta)=1+\frac{1}{2}=1.5.$

3. $a_1=3$ and $d=5$ imply that $S_{30}=\frac{30(2a_1+29d)}{2}=15(2(3)+29(5))=2265.$

4. $f(x)=5-2\cos^2(\frac{\pi x}{3})=5-2(\frac{1+\cos(2\pi x/3)}{2})=5-(1+\cos(\frac{2\pi}{3}x))$, so the period is $\frac{2\pi}{2\pi/3}=3.$

5. The radicand must be non-negative, so $8-2x-x^2\geq 0 \implies (x+4)(x-2)\leq 0 \implies -4\leq x\leq 2.$

6. $4!\times 3!=24\times 6=144.$

7. Let $\theta_a$ be the angle formed by the positive $x$-axis and the vector $\mathbf{a}$, and $\theta_b$ be the angle formed by the positive $x$-axis and the vector $\mathbf{b}$, Then, $\tan(\theta_a)=3$ and $\tan(\theta_b)=\frac{1}{2}$. Since
$\tan(\theta_a-\theta_b)=\frac{\tan(\theta_a)-\tan(\theta_b)}{1+\tan(\theta_a)\tan(\theta_b)}=\frac{3-1/2}{1+3(1/2)}=1$, so $\theta_a-\theta_b=45°$.

8. $y^2-25x^2=25 \implies \frac{y^2}{25}-\frac{x^2}{1}=1$, so the asymptotes are $y=\pm\frac{5}{1}x=\pm x$.

9. The product of $3-4i$ and $3+4i$ is $3^2+4^2=25$.

10. The domain of $f(x)$ satisfies $\frac{x}{10-x}\geq 0$, so $x(10-x)\geq 0$ where $x\neq 10$. Hence, $x(x-10)\leq 0$ where $x\neq 10$. Therefore, $0\leq x<10$.

11. $\frac{2x^2-3x+1}{x+1}=2x-5+\frac{6}{x+1}$, so the asymptotes are $x=-1$ or $y=2x-5$.

12. $y^2=4x$ has one input with two outputs. This is not a function.

13. $\frac{\sin(2\theta)}{1-\cos(2\theta)}=\frac{2\sin(\theta)\cos(\theta)}{1-(1-2\sin^2(\theta))}=\frac{2\sin(\theta)\cos(\theta)}{2\sin^2(\theta)}=\frac{\cos(\theta)}{\sin(\theta)}=\cot(\theta).$

14. Since $|x|\geq 0$ for $-2\leq x\leq 2$, the minimum value of $f(x)$ is 0.

15.

$$\begin{aligned}
2x^2-(5+y)x-(1+3y)&=0\\
(-(5+y))^2-4(-(1+3y))(2)&\geq 0\\
25+10y+y^2+8+24y&\geq 0\\
y^2+34y+33y&\geq 0\\
(y+1)(y+33)&\geq 0
\end{aligned}$$

Hence, $y\leq -33$ or $-1\leq y$.

# Minitest 9

**Minitest 9** includes

- Recursive Functions
- Linear Functions and Equations
- Space Geometry
- Quadratic Functions and Equations
- Probability
- Exponential Functions and Equations
- Statistics
- System of Equations
- Piecewise Functions and Equations
- Trigonometric Functions
- Parametric Functions

# MATHEMATICS LEVEL 2 MINITEST 9

REFERENCE INFORMATION

THE FOLLOWING INFORMATION IS FOR YOUR REFERERENCE IN ANSWERING SOME OF THE QUESTIONS IN THE TEST.

Volume of a right circular cone with radius $r$ and height $h$ : $V = \frac{1}{3}\pi r^2 h$

Volume of a sphere with radius $r$ : $V = \frac{4}{3}\pi r^3$

Volume of a pyramid with base area $B$ and height $h$ : $V = \frac{1}{3}Bh$

Surface Area of a sphere with radius $r$ : $S = 4\pi r^2$

DO NOT DETACH FROM THE BOOK.

## MATHEMATICS LEVEL 2 MINITEST 9

For each of the following problems, decide which is the BEST of the choices given. If the exact numerical value is not one of the choices, select the choice that best approximates this value. Then fill in the corresponding circle on the answer sheet.

Notes: (1) A scientific or graphing calculator will be necessary for answering some (but not all) of the questions in this test. For each question you will have to decide whether or not you should use a calculator.

(2) For some questions in this test you may have to decide whether your calculator should be in the radian mode or the degree mode.

(3) Figures that accompany problems in this test are intended to provide information useful in solving the problems. They are drawn as accurately as possible EXCEPT when it is stated in a specific problem that its figure is not drawn to scale. All figures lie in a plane unless otherwise indicated.

(4) Unless otherwise specified, the domain of any function $f$ is assumed to be the set of all real numbers $x$ for which $f(x)$ is a real number. The range of $f$ is assumed to be the set of all real numbers $f(x)$, where $x$ is in the domain of $f$.

(5) Reference information that may be useful in answering the questions in this test can be found on the page preceding Question 1.

USE THIS SPACE FOR SCRATCH WORK.

1. If $f(x+1) = x \cdot f(x)$ and $f(1) = 1$, then what is the value of $f(5)$?

(A) 120
(B) 30
(C) 24
(D) 12
(E) 6

**MATHEMATICS LEVEL 2 Test - *Continued***

USE THIS SPACE FOR SCRATCH WORK.

2. If $(x,y)$ is the point on a line $y = mx + b$, then what is the distance between two points $(x,y)$ and $(0,b)$?

(A) $|x|\sqrt{1+m^2}$
(B) $x\sqrt{1+m^2}$
(C) $\sqrt{x^2+m^2}$
(D) $b\sqrt{x^2+m^2}$
(E) $\sqrt{x^2+b^2}$

3. There is a line perpendicular to a plane. What is the set of all points that the distance from the plane is 5 and the distance from the line is 4?

(A) Two equal-sized circles.
(B) Two unequal-sized circles.
(C) A cylinder
(D) A circle
(E) A sphere

4. If $x(x+4) = 3t$ has only one real zero, then what is the value of $t$?

(A) $-\frac{4}{5}$
(B) $-\frac{3}{4}$
(C) $\frac{4}{3}$
(D) 0
(E) $-\frac{4}{3}$

**MATHEMATICS LEVEL 2 Test - *Continued***

USE THIS SPACE FOR SCRATCH WORK.

5. There are two fair, six-sided dice. One die has faces $1, 1, 2, 2, 2$ and 2. The other die has faces $1, 2, 2, 2, 2$ and 2. What is the probability that both 1 and 2 are shown when two dice are thrown?

(A) $\frac{28}{36}$
(B) $\frac{14}{36}$
(C) $\frac{10}{36}$
(D) $\frac{7}{36}$
(E) $\frac{4}{36}$

6. An initial deposit of \$100 is compounded yearly at an annual rate of 4.8% and no other deposit is allowed. How many years will it take the amount to exceed \$300?

(A) 3 years
(B) 15 years
(C) 22 years
(D) 24 years
(E) 42 years

## MATHEMATICS LEVEL 2 Test - *Continued*

USE THIS SPACE FOR SCRATCH WORK.

7. If there are two sets of data $A$ and $B$, i.e.,

$$A = \{0, 2, 4, 6, 8\}$$
$$B = \{3, 5, 7, 9, 11\}$$

which of the following must be true?

I. same mean

II. same standard deviation

III. same range

(A) I only
(B) II only
(C) I and II only
(D) II and III only
(E) I and III only

8. The total sum of **S**, **J**, and **I** is 9.9, where **S** is 4.2 more than **I**, and **J** is 0.45 more than **I**. If **S** and **J** are increased by 20% and 15%, respectively, what is the new sum of **S**, **J** and **I**?

(A) 10.51
(B) 10.60
(C) 11.23
(D) 11.42
(E) 13.66

**MATHEMATICS LEVEL 2 Test - *Continued***

USE THIS SPACE FOR SCRATCH WORK.

9. If $x$ is a real number in the interval $\frac{7}{8} \leq x \leq \frac{4}{3}$, then what is the probability that $x$ is greater than 1?

(A) $\frac{8}{11}$ (B) $\frac{3}{11}$ (C) $\frac{1}{3}$ (D) $\frac{1}{8}$ (E) $\frac{11}{24}$

10. The braking distance is the distance your car travels before stopping when the brakes are applied. The braking distance is directly proportional to the square of velocity. The braking distance is 22 feet when the velocity is 20 km per hour. What is the braking distance when the velocity is 60 km per hour?

(A) 38
(B) 66
(C) 161
(D) 198
(E) 256

11. If $f(x) = [x]$ represents the greatest integer less than or equal to $x$, and $g(x) = -\frac{1}{x}$, then $(f \circ g)(3) =$

(A) $-\frac{1}{4}$ (B) $-\frac{1}{3}$ (C) 1 (D) 0 (E) $-1$

## MATHEMATICS LEVEL 2 Test - *Continued*

USE THIS SPACE FOR SCRATCH WORK.

12. Corgi's blood pressure, $P(mmHg)$, at time $t$ seconds can be calculated by a function of the form

$$P(t) = 120 + 25\sin(2\pi t)$$

The blood pressure varies between the systolic pressure when the heart is contracting and the diastolic pressure when the heart is relaxing. Corgi's heart beats twice during one period of the blood pressure. Pulse is heart rate or the number of times the heart beats in one minute. What is the Corgi's pulse?

(A) 30
(B) 60
(C) 95
(D) 120
(E) 145

13. If $x(t) = t^2$ and $y(t) = t$ for non-negative real numbers $t$, then the graph of $(x(t), y(t)$ is best described by

(A) $y = x$
(B) $x = y^2$
(C) $x = \sqrt{y}$
(D) $y = \sqrt{x}$
(E) $y = -\sqrt{x}$

**MATHEMATICS LEVEL 2 Test - *Continued***

USE THIS SPACE FOR SCRATCH WORK.

14. Given a function $f(x) = ax^2 + bx$, where $a \neq 0, b \neq 0$, which of the following must be true?

I. The range of $f(x)$ is all real numbers.

II. There are two distinct $x$-intercepts.

III. The $y$-intercept is 0.

(A) I only
(B) II only
(C) II and III only
(D) III only
(E) I, II, and III

15. Which of the following function satisfies the following condition : if $a < b$, then $f(a) \leq f(b)$ for all real numbers $a$, $b$?

(A) $y = e^{-x}$
(B) $y = -\sqrt{x}$
(C) $y = |x|$
(D) $y = e^{x}$
(E) $y = -\log(x)$

## Answerkey to Minitest 9

1. (C)

2. (A)

3. (A)

4. (E)

5. (B)

6. (D)

7. (D)

8. (D)

9. (A)

10. (D)

11. (E)

12. (D)

13. (D)

14. (C)

15. (D)

## Solution

1. $f(2) = 1f(1) = 1$, $f(3) = 2f(2) = 2$, $f(4) = 3f(3) = 6$, and $f(5) = 4f(4) = 24$.

2. $\sqrt{(x-0)^2+(y-b)^2} = \sqrt{x^2+(mx+b-b)^2} = \sqrt{x^2+m^2x^2} = \sqrt{x^2(1+m^2)} = \sqrt{x^2}\sqrt{1+m^2} = |x|\sqrt{1+m^2}$.

3. Think about the plane as the flat ground. Then, the set of points 5 units away from the plane should be the plane 5 units above or below the ground. The set of points 4 units from the line in space is the cylinder of radius of 4 with infinite height. The intersection points between the two sets must be two equal-sized circles.

4. $x^2+4x-3t = 0$ has one zero if $4^2-4(-3t) = 0 \implies 16+12t = 0 \implies t = -\frac{16}{12} = -\frac{4}{3}$.

5. Let's case-enumerate. The first case of 1 and 2 is $\frac{2}{6} \times \frac{5}{6} = \frac{10}{36}$. The second case of 2 and 1 is $\frac{4}{6} \times \frac{1}{6} = \frac{4}{36}$. Hence, the total probability of 1 and 2 showing up when two dice are thrown is $\frac{14}{36}$.

6. $100 \times (1+0.048)^t > 300$, so $t = 24$.

7. From $A$ to $B$, 3 is added to each data value, so the mean changes, but standard deviation or range do not change.

8. Let's solve this by system of equations with three variables.

$$\begin{cases} S+J+I = 9.9 \\ S = I+4.2 \\ J = I+0.45 \end{cases} \implies 3I = 5.25$$

$$\implies I = 1.75$$
$$\implies J = 2.2$$
$$\implies S = 5.95$$

Hence, the new sum $1.2S+1.15J+I$ equals $1.75(1.2)+2.2(1.15)+5.95 = 11.42$.

9. $\frac{4/3-1}{4/3-7/8} = \frac{8}{11}$.

10. Let the braking distance be $x$ and the velocity be $y$. Then, $x = k \times y^2$. Since $22 = k(20)^2$, $y = \frac{22}{400}(60)^2 = 198$.

11. $f(g(3)) = f(-\frac{1}{3}) = \lfloor -\frac{1}{3} \rfloor = -1$.

12. Corgi's heart beats twice per $\frac{2\pi}{2\pi}(=1)$ second. By the definition of pulse in the question, it must be the number of heart beats in one minute, where Corgi's heart beats twice per second. Hence, pulse must be 120.

13. Since $y = t$, then $x = t^2 = y^2$. The graph of $x = y^2$ is a full parabola, but the graph of $x(t) = t^2, y = t$ for $t \geq 0$ is the portion of parabola above the $x$-axis. Hence, the correct relationship must be $y = \sqrt{x}$.

14. $f(x) = ax^2 + bx = x(ax + b)$, so the graph is a parabola with the $x$-intercepts of 0 and $-\dfrac{b}{a}$. Also, $f(0) = 0$. Thus, II and III are true.

15. If $a < b$, then $f(a) \leq f(b)$ for $y = e^x$. The question asks us to find an increasing function, and it is the only increasing function out of all answer choices.

# Minitest 10

**Minitest 10** includes

- Number Theory
- Permutations and Combinations
- Linear Functions and Equations
- Conic Sections
- Sequence and Series
- Plane Geometry
- Logic
- Probability
- Exponential Functions and Equations
- Quadratic Functions and Equations

# MATHEMATICS LEVEL 2 MINITEST 10

REFERENCE INFORMATION

THE FOLLOWING INFORMATION IS FOR YOUR REFERERENCE IN ANSWERING SOME OF THE QUESTIONS IN THE TEST.

Volume of a right circular cone with radius $r$ and height $h$ : $V = \frac{1}{3}\pi r^2 h$

Volume of a sphere with radius $r$ : $V = \frac{4}{3}\pi r^3$

Volume of a pyramid with base area $B$ and height $h$ : $V = \frac{1}{3}Bh$

Surface Area of a sphere with radius $r$ : $S = 4\pi r^2$

DO NOT DETACH FROM THE BOOK.

## MATHEMATICS LEVEL 2 MINITEST 10

For each of the following problems, decide which is the BEST of the choices given. If the exact numerical value is not one of the choices, select the choice that best approximates this value. Then fill in the corresponding circle on the answer sheet.

Notes: (1) A scientific or graphing calculator will be necessary for answering some (but not all) of the questions in this test. For each question you will have to decide whether or not you should use a calculator.

(2) For some questions in this test you may have to decide whether your calculator should be in the radian mode or the degree mode.

(3) Figures that accompany problems in this test are intended to provide information useful in solving the problems. They are drawn as accurately as possible EXCEPT when it is stated in a specific problem that its figure is not drawn to scale. All figures lie in a plane unless otherwise indicated.

(4) Unless otherwise specified, the domain of any function $f$ is assumed to be the set of all real numbers $x$ for which $f(x)$ is a real number. The range of $f$ is assumed to be the set of all real numbers $f(x)$, where $x$ is in the domain of $f$.

(5) Reference information that may be useful in answering the questions in this test can be found on the page preceding Question 1.

1. Let $x$ be the smallest positive integer such that $2015+x$ is a perfect square. Let $y$ be the smallest positive integer such that $2015-y$ is a perfect square. What is the value of $x+y$?

(A) 85
(B) 87
(C) 89
(D) 91
(E) 93

USE THIS SPACE FOR SCRATCH WORK.

## MATHEMATICS LEVEL 2 Daily Test - *Continued*

USE THIS SPACE FOR SCRATCH WORK.

2. If $A$ and $B$ are digits such that $123 \times 4A6 = 5B548$, the value of $A \times B$, the product of the two missing digits, equals

(A) 50
(B) 52
(C) 54
(D) 56
(E) 58

3. In a liberal arts highschool called Awesomeness, there are twenty teachers, ten of whom teach mathematics, eight of whom teach literature and six of whom teach science. Two teach both mathematics and literature, but none teach both literature and science. How many teachers teach both mathematics and science?

(A) 1
(B) 2
(C) 3
(D) 4
(E) 5

4. If 21 is expressed as a sum of $n$ consecutive positive integers, what is the greatest possible value of $n$?

(A) 6
(B) 7
(C) 8
(D) 9
(E) 10

**MATHEMATICS LEVEL 2 Daily Test - *Continued***

USE THIS SPACE FOR SCRATCH WORK.

5. The sum of all of the four-digit numbers whose digits arrangements are permutations of 1, 2, 3, and 4, is equal to

(A) $60,000$
(B) $66,000$
(C) $66,600$
(D) $66,660$
(E) $66,666$

6. If the point $(8,k)$ in the first quadrant has the same distance from the point $(0,4)$ as it is from the $x$-axis, what is the value of $k$?

(A) 10
(B) 12
(C) 14
(D) 16
(E) 18

7. The infinite geometric series $4+2\sqrt{2}+2+\sqrt{2}+\ldots$, each term found by dividing the previous term by $\sqrt{2}$, in simplest radical form, has the summation value of $m+n\sqrt{2}$, where $m+n=$

(A) 12
(B) 10
(C) 8
(D) 6
(E) 4

**MATHEMATICS LEVEL 2 Daily Test - *Continued***

USE THIS SPACE FOR SCRATCH WORK.

8. 99 precinct NYPD had a tactical training once a year, where ten people, including perpetrators and officers, were positioned in a closed room so that no two of them stood the same distance apart. Each person aimed his or her paintball pistol at his or her closest opponent, and at the signal everyone fired. What is the maximum number of times Jake Peralta, a 'detective/genius,' could have been hit?

(A) 4
(B) 5
(C) 6
(D) 7
(E) 8

9. All points with coordinates $(x, y)$, equidistant from the points $(1,3)$ and $(7,11)$ lie on a single line. If the equation of the line is written in the form $y = mx + b$, what is the value of $b$?

(A) 10 (B) 9 (C) 8 (D) 7 (E) 6

10. The units digit of the product $2^{2015} \times 7^{2015}$ equals

(A) 1 (B) 2 (C) 4 (D) 6 (E) 7

## MATHEMATICS LEVEL 2 Daily Test - *Continued*

USE THIS SPACE FOR SCRATCH WORK.

11. A mathematical statement "If P is true, then Q is true" is equivalent to

(A) If P is false, then Q is true.
(B) If P is true, then Q is false.
(C) If P is false, then Q is false.
(D) If Q is true, then P is true.
(E) If Q is false, then P is false.

12. From a deck of 52 cards, what is the probability of drawing three black cards in a row if each drawn card is not returned to the deck?

(A) $\frac{1}{8}$ (B) $\frac{4}{33}$ (C) $\frac{2}{17}$ (D) $\frac{5}{41}$ (E) $\frac{3}{25}$

13. Which of the following is the value of $p$ if $2^3 + 2^4 = (p-2)2^4$?

(A) $-\frac{1}{2}$ (B) 10 (C) 3 (D) $\frac{7}{2}$ (E) $\frac{5}{2}$

## MATHEMATICS LEVEL 2 Daily Test - *Continued*

USE THIS SPACE FOR SCRATCH WORK.

14. Arithmetic sequence is a sequence of numerical terms, subsequent value of which is obtained by adding or subtracting a constant, known as the common difference, from the previous term. If the sum of the first 50 terms of an arithmetic sequence is 100, where the common difference is 2, what is the value of the first term?

(A) 47
(B) $-47$
(C) 48
(D) $-51$
(E) $-48$

15. If $f(x) = (x-1)^2 + (x+1)^2$ for all real numbers $x$, which of the following are true?

I. $f(x) = f(-x)$

II. $f(x) = f(x+1)$

III. $f(x) = |f(x)|$

(A) I only
(B) II and III only
(C) III only
(D) I and III only
(E) I, II, and III

## Answerkey to Minitest 10

1. (C)
2. (D)
3. (B)
4. (A)
5. (D)
6. (A)
7. (A)
8. (B)
9. (A)
10. (C)
11. (E)
12. (C)
13. (D)
14. (B)
15. (D)

## Solution

1. Since $45^2 = 2025$, $2015+n = 2025 \implies n = 10$. Similarly, $44^2 = 1936$, so $2015-m = 1936 \implies m = 79$. Therefore, $m+n = 89$.

2. Since $3A+3 = 14$ or $24$, we get $3A = 11$ or $21$. Hence, $A = 7$ and $B = 8$, Therefore, $A \times B = 56$.

3. Use the principle of inclusion and exclusion. In Awesomenass, $10+8+6-x-2 = 20$ where $x$ is the number of teachers teaching both math and science. Hence, $x = 2$.

4. $1+2+3+4+5+6 = 21$, so $n = 6$ is the greatest possible value of $n$.

5. $1234+1324+1342+\cdots+4321 = (1+2+3+4)\times 6+(10+20+30+40)\times 6+(100+200+300+400)\times 6+(1000+2000+3000+4000)\times 6 = 66,660$.

6. $\sqrt{8^2+(k-4)^2} = k \implies 64+k^2-8k+16 = k^2 \implies 8k = 80 \implies k = 10$.

7. $4+2\sqrt{2}+2+\sqrt{2}+\cdots = \dfrac{4}{1-1/\sqrt{2}} = 4(2+\sqrt{2}) = 8+4\sqrt{2}$, so $m+n = 12$.

8. Since at most 5 angles greater than $60^\circ$ make $360^\circ$, so Peralta could have been hit by at most 5 people.

9. $\sqrt{(x-1)^2+(y-3)^2} = \sqrt{(x-7)^2+(y-11)^2}$, so $3x+4y = 40 \implies 4y = -3x+40 \implies y = -\dfrac{3}{4}x+10$. The value of $b$ is 10.

10. $14^{2015} \equiv 4^{2015} \equiv 2^{4030} \equiv 4 \mod 10$.

11. "If P is true, then Q is true" is equivalent to "If Q is false, then P is false."

12. $\dfrac{26}{52}\times\dfrac{25}{51}\times\dfrac{24}{50} = \dfrac{6}{51} = \dfrac{2}{17}$.

13. $2^3(1+2) = (2p-4)2^3 \implies 3 = 2p-4 \implies 2p = 7 \implies p = \dfrac{7}{2}$.

14.

$$\frac{50(2a_1+49(2))}{2} = 100$$
$$2a_1+98 = 4$$
$$2a_1 = -94$$
$$a_1 = -47$$

15. If $f(x) = (x-1)^2+(x+1)^2 = x^2-2x+1+x^2+2x+1 = 2x^2+2$. Since the function is even, $f(x) = f(-x)$. Also, since $f(x) \geq 2 > 0$, then $f(x) = |f(x)|$.

# Minitest 11

**Minitest 11** includes

- Trigonometric Functions
- Number Theory
- Statistics
- Counting
- Plane Geometry
- Linear Functions and Equations
- Space Geometry
- Matrices
- Quadratic Functions and Equations
- Rational Functions and Equations
- Parametric Functions

# MATHEMATICS LEVEL 2 MINITEST 11

REFERENCE INFORMATION

THE FOLLOWING INFORMATION IS FOR YOUR REFERERENCE IN ANSWERING SOME OF THE QUESTIONS IN THE TEST.

Volume of a right circular cone with radius $r$ and height $h$ : $V = \frac{1}{3}\pi r^2 h$

Volume of a sphere with radius $r$ : $V = \frac{4}{3}\pi r^3$

Volume of a pyramid with base area $B$ and height $h$ : $V = \frac{1}{3}Bh$

Surface Area of a sphere with radius $r$ : $S = 4\pi r^2$

DO NOT DETACH FROM THE BOOK.

## MATHEMATICS LEVEL 2 MINITEST 11

For each of the following problems, decide which is the BEST of the choices given. If the exact numerical value is not one of the choices, select the choice that best approximates this value. Then fill in the corresponding circle on the answer sheet.

Notes: (1) A scientific or graphing calculator will be necessary for answering some (but not all) of the questions in this test. For each question you will have to decide whether or not you should use a calculator.

(2) For some questions in this test you may have to decide whether your calculator should be in the radian mode or the degree mode.

(3) Figures that accompany problems in this test are intended to provide information useful in solving the problems. They are drawn as accurately as possible EXCEPT when it is stated in a specific problem that its figure is not drawn to scale. All figures lie in a plane unless otherwise indicated.

(4) Unless otherwise specified, the domain of any function $f$ is assumed to be the set of all real numbers $x$ for which $f(x)$ is a real number. The range of $f$ is assumed to be the set of all real numbers $f(x)$, where $x$ is in the domain of $f$.

(5) Reference information that may be useful in answering the questions in this test can be found on the page preceding Question 1.

USE THIS SPACE FOR SCRATCH WORK.

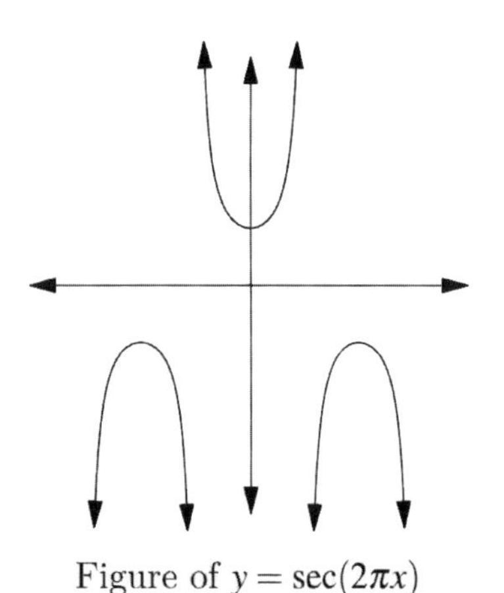

Figure of $y = \sec(2\pi x)$

1. What is the period of $f(x) = \sec(2\pi x)$?

(A) 1 (B) 2 (C) 3 (D) 4 (E) $\frac{1}{2}$

**MATHEMATICS LEVEL 2 Test - *Continued***

USE THIS SPACE FOR SCRATCH WORK.

2. What is the number of all positive integral values of $n$ for which $\frac{n+6}{n}$ is an integer?

(A) 3
(B) 4
(C) 5
(D) 6
(E) None of the above

3. A fifth number, $n$ , is added to the set $\{3, 6, 9, 10\}$ to make the mean of the set of five numbers equal to its median. The number of possible values of $n$ is

(A) 1
(B) 2
(C) 3
(D) 4
(E) more than 4

4. How many fractions in the form $\frac{n}{99}$, where $0 < n < 99$, are in lowest terms?

(A) 38
(B) 48
(C) 50
(D) 60
(E) 98

**MATHEMATICS LEVEL 2 Test - *Continued***

USE THIS SPACE FOR SCRATCH WORK.

5. How many positive integers, not exceeding 100, are multiples of 2 or 3 but not 4?

(A) 25 (B) 42 (C) 50 (D) 67 (E) 88

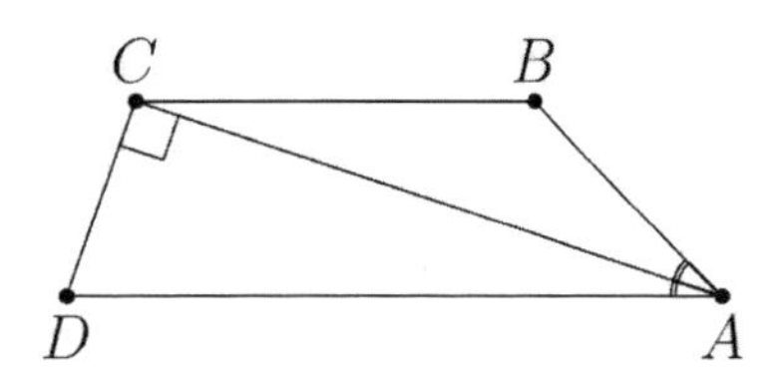


Figure NOT drawn to scale.

6. *ABCD* is a trapezoid in which $\overline{AD}$ is parallel to $\overline{BC}$. Given $\overline{AC} \perp \overline{CD}$, $\overline{AC}$ bisects angle $\angle BAD$, if the area of *ABCD* is 42, then the area of triangle *ACD* is

(A) 14 (B) 21 (C) 28 (D) 35 (E) 42

7. What is the number of distinct counting numbers that leave a remainder of 5 when divided into 47?

(A) 4 (B) 5 (C) 6 (D) 7 (E) 8

**MATHEMATICS LEVEL 2 Test - *Continued***

USE THIS SPACE FOR SCRATCH WORK.

8. Bob bought 4 different items at a grocery store. He brought 3 identical bags, and handed them to the cashier. How many ways are there for the cashier to put the items Bob bought in the 3 identical bags, assuming he might leave some of the bags empty?

(A) 8
(B) 10
(C) 12
(D) 13
(E) 14

9. If $f(x) = 2x + 3$, then $f(f(x)) = ax + b$. Which of the following is $a + b$?

(A) 4
(B) 6
(C) 9
(D) 13
(E) 36

10. Rectangle $ABCD$ is the base of pyramid $PABCD$. If $AB = 3$, $BC = 2$, $\overline{PA} \perp \overline{AD}$, $\overline{PA} \perp \overline{AB}$, and $PC = 5$, then what is the volume of the pyramid $PABCD$?

(A) $2\sqrt{3}$
(B) $4\sqrt{3}$
(C) 6
(D) 12
(E) 24

**MATHEMATICS LEVEL 2 Test - *Continued***

USE THIS SPACE FOR SCRATCH WORK.

11. Find the value of $x$ if the determinant of the matrix $A$ is $2x-4$ where

$$A = \begin{bmatrix} x-2 & 0 \\ 0 & x-3 \end{bmatrix}$$

(A) $x=2$ or $x=5$
(B) $x=2$ or $x=3$
(C) $x=1$ or $x=-1$
(D) $x=-2$ or $x=1$
(E) $x=0$ or $x=2$

12. Which of the following represents all possible values of $q$ in the equation $4x^2+qx+12$ so that the roots $r_1$ and $r_2$ satisfy $\frac{r_1}{r_2}=3$?

(A) $0, 16, -16$
(B) $-16$
(C) $16, -16$
(D) $0$
(E) $16$

13. If $f(x)=\frac{x}{x+1}$, what is $f(2x)$ in terms of $f(x)$?

(A) $\frac{f(x)}{1-f(x)}$
(B) $\frac{f(x)}{1+2f(x)}$
(C) $\frac{2f(x)}{1-f(x)}$
(D) $\frac{2f(x)}{1+f(x)}$
(E) $\frac{2f(x)}{2f(x)+1}$

**MATHEMATICS LEVEL 2 Test - *Continued***

USE THIS SPACE FOR SCRATCH WORK.

14. Which of the following is the shape of the graph of $(x(t), y(t)) = \left(t + \frac{1}{t}, t - \frac{1}{t}\right)$?

(A) an ellipse
(B) a circle
(C) a straight line
(D) a parabola
(E) a hyperbola

15. Which of the following has a root $12 - 13i$, where $i = \sqrt{-1}$?

(A) $x^2 - 24x - 313 = 0$
(B) $x^2 - 24x + 313 = 0$
(C) $x^2 + 24x - 25 = 0$
(D) $x^2 + 26x + 313 = 0$
(E) $x^2 + 24x + 25 = 0$

## Answerkey to Minitest 11

1. (A)
2. (B)
3. (C)
4. (D)
5. (B)
6. (C)
7. (B)
8. (E)
9. (D)
10. (B)
11. (A)
12. (C)
13. (D)
14. (E)
15. (B)

## Solution

1. The period of $y = \sec(2\pi x)$ is $\dfrac{2\pi}{2\pi} = 1$.

2. $\dfrac{n+6}{n} = 1 + \dfrac{6}{n}$ is integer if and only if $n$ is a factor of 6. Hence, the positive factor of 6 is $1, 2, 3, 6$.

3. Since $n + 28 = 5n, 30, 45$, then either $n = 7$, $n = 2$, or $n = 17$. There are three possible values of $n$.

4. $\lfloor \frac{98}{3} \rfloor + \lfloor \frac{98}{11} \rfloor - \lfloor \frac{98}{33} \rfloor = 38$. Hence, $98 - 38 = 60$.

5. $\lfloor \frac{100}{2} \rfloor + \lfloor \frac{100}{3} \rfloor - \lfloor \frac{100}{6} \rfloor = 50 + 33 - 16 = 67$.

6.

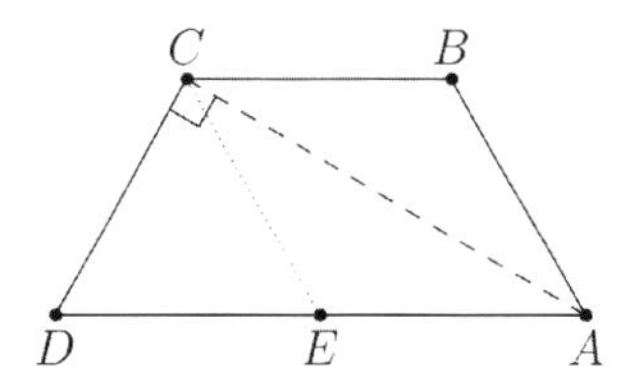


Since $\overline{BC}$ is parallel to $\overline{AD}$, $m\angle CAD = m\angle ACB$, so $\triangle ABC$ is isosceles. Also, $\triangle AEC$ is isosceles with $AE = EC$. On the other hand, $\triangle CDE$ is isosceles with $DE = CE$. Hence, all three triangles have the same area. Therefore, $\triangle ACD$ has the area of $42 \times \dfrac{2}{3} = 28$.

7. $47 = nk + 5$, so $nk = 42$. Hence, the possible values of $n > 5$, must be $42, 21, 14, 7, 6$. There are five values for $n$.

8. Let $x + y + z = 4$ where $x \geq y \geq z$. Then, $(x, y, z) = (4,0,0), (3,1,0), (2,2,0), (2,1,1)$. Hence, the total possibilities are $\binom{4}{4} + \binom{4}{3} \times \binom{1}{1} + \dfrac{\binom{4}{2} \times \binom{2}{2}}{2!} + \dfrac{\binom{4}{2} \times \binom{2}{1} \times \binom{1}{1}}{2!} = 14$.

9. $f(f(x)) = 2(2x+3) + 3 = 4x + 9 = ax + b$, so $a + b = 4 + 9 = 13$.

10. Since $AC = \sqrt{3^2 + 2^2} = \sqrt{13}$, then $PA = \sqrt{25 - 13} = \sqrt{12}$. Hence, the volume of the pyramid is $\dfrac{1}{3} \times 2 \times 3 \times \sqrt{12} = 4\sqrt{3}$.

11. The determinant of $A$ equals $(x-2)(x-3) - (0)(0) = 2x - 4$, so $x = 2$ or $x = 5$.

12. Since $r_1 r_2 = 3$ and $\dfrac{r_1}{r_2} = 3$, then $r_2 = 1$, and $(r_1, r_2) = (3, 1), (-3, -1)$. Hence, by Vieta's formula, $-\dfrac{q}{4} = 4$ or $-4$. Hence, $q = \pm 16$.

13. $\dfrac{2f(x)}{1 + f(x)} = \dfrac{2x}{1 + 2x} = f(2x)$.

14. $x^2 = t^2 + 2 + \dfrac{1}{t^2}$ and $y^2 = t^2 - 2 + \dfrac{1}{t^2}$, so $x^2 - y^2 = 4$, which is a hyperbola.

15. $x - 12 = -13i \implies (x-12)^2 = -169 \implies x^2 - 24x + 144 = -169 \implies x^2 - 24x + 313 = 0$

# Minitest 12

**Minitest 12** includes

- Inverse Functions
- Sequence and Series
- Space Geometry
- Linear Functions and Equations
- Statistics
- Counting
- Logarithm
- Probability
- Plane Geometry
- Trigonometric Functions
- Polynomial Functions and Equations
- Polar Coordinates

# MATHEMATICS LEVEL 2 MINITEST 12

REFERENCE INFORMATION

THE FOLLOWING INFORMATION IS FOR YOUR REFERERENCE IN ANSWERING SOME OF THE QUESTIONS IN THE TEST.

Volume of a right circular cone with radius $r$ and height $h$ : $V = \frac{1}{3}\pi r^2 h$

Volume of a sphere with radius $r$ : $V = \frac{4}{3}\pi r^3$

Volume of a pyramid with base area $B$ and height $h$ : $V = \frac{1}{3}Bh$

Surface Area of a sphere with radius $r$ : $S = 4\pi r^2$

DO NOT DETACH FROM THE BOOK.

## MATHEMATICS LEVEL 2 MINITEST 12

For each of the following problems, decide which is the BEST of the choices given. If the exact numerical value is not one of the choices, select the choice that best approximates this value. Then fill in the corresponding circle on the answer sheet.

Notes: (1) A scientific or graphing calculator will be necessary for answering some (but not all) of the questions in this test. For each question you will have to decide whether or not you should use a calculator.

(2) For some questions in this test you may have to decide whether your calculator should be in the radian mode or the degree mode.

(3) Figures that accompany problems in this test are intended to provide information useful in solving the problems. They are drawn as accurately as possible EXCEPT when it is stated in a specific problem that its figure is not drawn to scale. All figures lie in a plane unless otherwise indicated.

(4) Unless otherwise specified, the domain of any function $f$ is assumed to be the set of all real numbers $x$ for which $f(x)$ is a real number. The range of $f$ is assumed to be the set of all real numbers $f(x)$, where $x$ is in the domain of $f$.

(5) Reference information that may be useful in answering the questions in this test can be found on the page preceding Question 1.

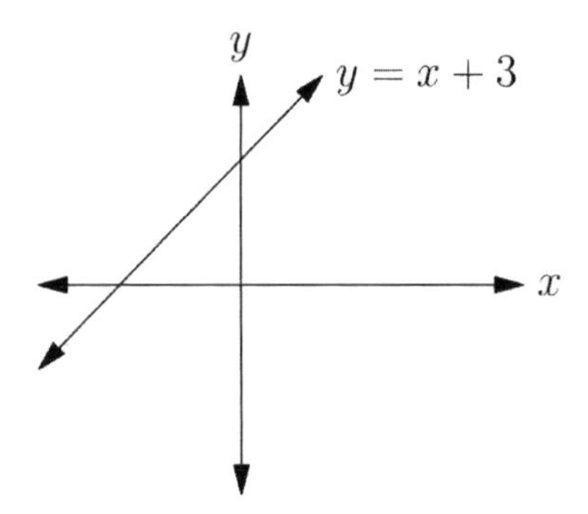


USE THIS SPACE FOR SCRATCH WORK.

1. If $f(x) = x + 3$, then $f^{-1}(2) =$

(A) $-1$
(B) $-2$
(C) $-3$
(D) $-4$
(E) $-5$

**MATHEMATICS LEVEL 2 Test - *Continued***

USE THIS SPACE FOR SCRATCH WORK.

2. The 7th term of a geometric sequence defined as 243, 81, 27, $\cdots$ is

(A) $-\frac{1}{27}$

(B) $\frac{1}{9}$

(C) $\frac{1}{3}$

(D) 1

(E) 3

3. What is the ratio of the surface area of two distinct spheres each of which has the radius of $2r$ and $3r$?

(A) $\frac{2}{3}$ (B) $\frac{4}{9}$ (C) $\frac{8}{37}$ (D) $\frac{7}{18}$ (E) $\frac{4}{27}$

4. If $g(x)=3x$, $f(x)=2$, and $h(x)=-2x+3$, then $h(f(g(1)))$?

(A) $-3$

(B) $-1$

(C) 0

(D) 1

(E) 3

**MATHEMATICS LEVEL 2 Test - *Continued***

USE THIS SPACE FOR SCRATCH WORK.

5. Given a recursive sequence with $t_1 = 0.5$ and $t_n = 2t_{n-1}$, which of the following is equal to $t_k$?

(A) $2^{k-2}$
(B) $2^{k-1}$
(C) $2^k$
(D) $2^{k+1}$
(E) $2^{k+2}$

6. What is the arithmetic mean of integers from 1 to 137, inclusive?

(A) 68
(B) 69
(C) 71
(D) 137
(E) 138

7. In a box, there are five pairs of black, white, red, yellow, and blue socks. At least how many socks should be picked to guarantee a pair of the same color?

(A) 2 (B) 3 (C) 4 (D) 5 (E) 6

**MATHEMATICS LEVEL 2 Test - *Continued***

USE THIS SPACE FOR SCRATCH WORK.

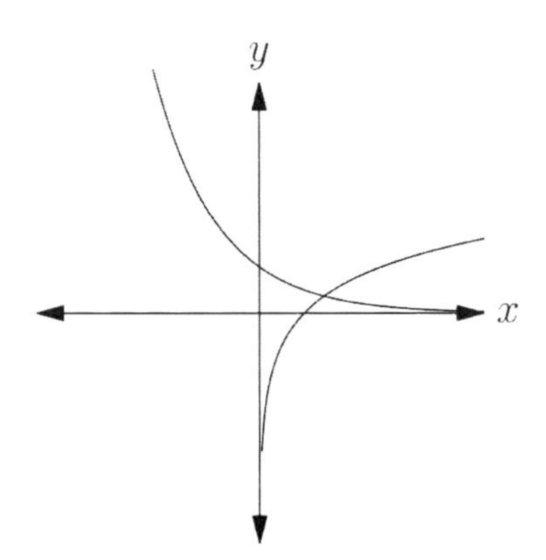


8. If $\log_3(x) = 2^{-x}$, which of the following is the value of $\log_5(x)$?

(A) 0.17
(B) 0.24
(C) 0.29
(D) 0.39
(E) 0.47

9. Let $x$ be an integer such that $10 \leq x \leq 22$. What is the probability that $x$ is an odd integer?

(A) $\frac{6}{13}$ (B) $\frac{7}{13}$ (C) $\frac{1}{2}$ (D) $\frac{5}{12}$ (E) $\frac{1}{3}$

**MATHEMATICS LEVEL 2 Test - *Continued***

USE THIS SPACE FOR SCRATCH WORK.

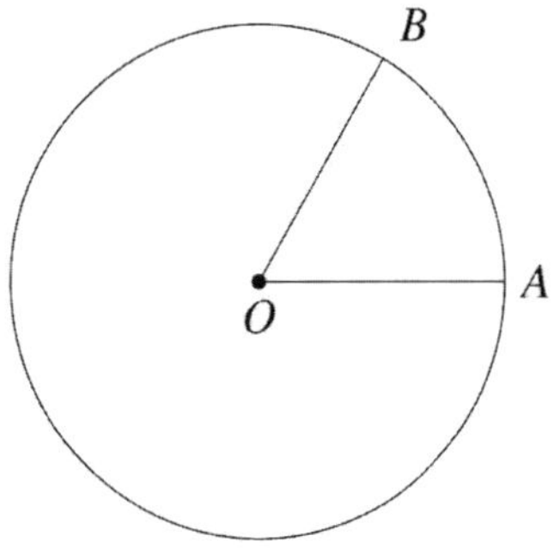

B

O

A

10. Given a circle, if the length of the chord $\overline{AB}$ of $60^\circ$ arc is 2 inches long, what is the area of the circle?

(A) $\frac{\pi}{2}$ (B) $\pi$ (C) $\frac{3\pi}{2}$ (D) $2\pi$ (E) $4\pi$

11. In the middle of pacific ocean, there is a famous spot where whales appear periodically. The number of whales, $N$, can be modeled by

$$N(t) = 30 + 15\sin(2\pi t)$$

where $t$ is the number of days passed since the first day of observation. How many times would you spot whales at its peak over the course of 5 weeks since the beginning of the observation?

(A) 20
(B) 25
(C) 30
(D) 35
(E) 40

## MATHEMATICS LEVEL 2 Test - *Continued*

12. If $f(x) = ax^4 - bx^3$ where $a, b \neq 0$, which of the following must be true?

USE THIS SPACE FOR SCRATCH WORK.

I. $f(x)$ has at least one repeated root.

II. $f(1) > 0$.

III. Function $f(x)$ is onto the set of real numbers.

(A) I only
(B) I and II
(C) II and III
(D) I and III
(E) I, II, and III

13. Which of the following is true for $f(x) = x^3 + x + 3$ and $g(x) = x^2 - 3x + 2$?

(A) $g(x) = 0$ does not have a real root.
(B) The graph of $f(x) + g(x)$ is concave up and its vertex is above the $x$-axis. Therefore, $f(x) + g(x)$ has no $x$-intercept.
(C) $f(x) = g(x)$ has one real root.
(D) The graph of $f(x)$ is concave up and the $x$-coordinate of the vertex is negative. Hence, $f(x)$ has no real root.
(E) There is no real solution satisfying $f(x) = g(x)$.

**MATHEMATICS LEVEL 2 Test - *Continued***

USE THIS SPACE FOR SCRATCH WORK.

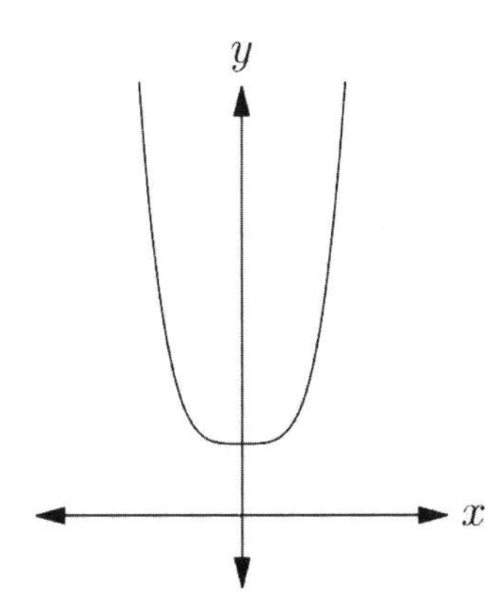


Figure of $y = x^4 + 1$

14. If $f(x) = x^4 + 1$ for $x \geq 0$, then $f^{-1}(x) =$

(A) $\sqrt[4]{x} + 1$
(B) $x^4 - 1$
(C) $\sqrt[4]{x-1}$
(D) $\sqrt[4]{x+1}$
(E) $\sqrt[4]{x} - 1$

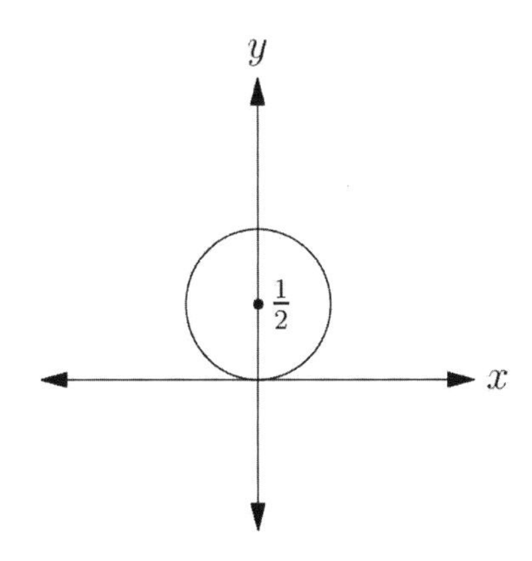


15. Which of the following can be a polar coordinate of $r = \sin(\theta)$?

(A) $\left(1, \frac{\pi}{2}\right)$
(B) $\left(1, \frac{\pi}{4}\right)$
(C) $\left(\sqrt{3}, \frac{2\pi}{3}\right)$
(D) $\left(\sqrt{2}, \frac{\pi}{3}\right)$
(E) $\left(1, \frac{2\pi}{3}\right)$

## Answerkety to Minitest 12

1. (A)

2. (C)

3. (B)

4. (B)

5. (A)

6. (B)

7. (E)

8. (B)

9. (A)

10. (E)

11. (D)

12. (A)

13. (C)

14. (C)

15. (A)

## Solution

1. Let $f^{-1}(2) = x$. Then $2 = f(x) = x+3$, so $x = -1$.

2. The sequence in the question is a geometric sequence with the common ratio of $\frac{1}{3}$. Hence, the 7th term is $243 \times (\frac{1}{3})^6 = \frac{1}{3}$.

3. Sphere are always similar. If the length ratio between two spheres is $2:3$, then the area ratio must be $4:9$ or $\frac{4}{9}$.

4. $h(f(g(1))) = h(f(3)) = h(2) = -2(2)+3 = -1$.

5. Substituting $k = 1$, $t_k = 2^{k-2}$ is the only expression satisfying $t_1 = 0.5$.

6. $\frac{1+2+3+\cdots+137}{137} = \frac{137(1+137)}{2(137)} = \frac{138}{2} = 69$.

7. Imagine the worst. Every time when you pick a sock, suppose different colored one comes out. However, at the sixth pick, whichever sock you pick, you will get a pair. This is called Pigeonhole principle.

8. Use a graphic calculator to find out the $x$-coordinate of the intersection point, roughly 1.4818. Hence, $\log_5(1.4818) \approx 0.24$.

9. Out of $13(= 22-10+1)$ integers, there are $6(= 11, 13, 15, 17, 19, 21)$ odd integers, so the probability is $\frac{6}{13}$.

10. Since $\triangle BOA$ is an equilateral triangle, the radius must be 2. Hence, the area of a circle with the radius is $(2)^2\pi = 4\pi$.

11. Since the period of $N(t)$ is $\frac{2\pi}{2\pi} = 1$, there is 1 peak per day. Therefore, there are 35 peaks over 35 days(=5 weeks).

12. $f(x) = x^3(ax-b)$ implies that the function has three repeated roots. II and III are not implied by the quartic function, so only I is true.

13.

- $y = g(x)$ has two real $x$-intercepts.
- The graph of $f(x)+g(x)$ is cubic graph, meaning that some portions of the graph are concave up while others are not.
- The graph of $f(x)$, a cubic function, cannot be concave up nor concave down over all $x$.
- $f(x) = g(x)$ has one real solution, so (C) must be true.

14. Since $f(x) = x^4 + 1$ and $f(f^{-1}(x)) = x$, then

$$\begin{aligned}(f^{-1}(x))^4 + 1 &= x \\ (f^{-1}(x))^4 &= x - 1 \\ f^{-1}(x) &= \sqrt[4]{x-1}\end{aligned}$$

15. As one can check from the figure in the question, $r = \sin(\theta)$ turns into

$$\begin{aligned}r &= \sin(\theta) \\ r^2 &= r\sin(\theta) \\ x^2 + y^2 &= y \\ x^2 + (y^2 - y + \frac{1}{4}) &= \frac{1}{4} \\ x^2 + (y - \frac{1}{2})^2 &= (\frac{1}{2})^2\end{aligned}$$

In order to check whether polar coordinates satisfy the given polar equation, we simply substitute the answer choices in the equation, i.e., $1 = \sin(\frac{\pi}{2})$, so (A) is true.

# Minitest 13

**Minitest 13** includes

- Quadratic Functions and Equations
- Polynomial Functions and Equations
- Inverse Functions
- Trigonometric Identities
- Piecewise Functions
- Exponential Functions and Equations
- Space Geometry
- Combinations and Permutations
- Rational Functions and Equations
- Probability
- Logarithm
- Matrices
- Trigonometric Equations

# MATHEMATICS LEVEL 2 MINITEST 13

REFERENCE INFORMATION

THE FOLLOWING INFORMATION IS FOR YOUR REFERERENCE IN ANSWERING SOME OF THE QUESTIONS IN THE TEST.

Volume of a right circular cone with radius $r$ and height $h$ : $V = \frac{1}{3}\pi r^2 h$

Volume of a sphere with radius $r$ : $V = \frac{4}{3}\pi r^3$

Volume of a pyramid with base area $B$ and height $h$ : $V = \frac{1}{3}Bh$

Surface Area of a sphere with radius $r$ : $S = 4\pi r^2$

DO NOT DETACH FROM THE BOOK.

## MATHEMATICS LEVEL 2 MINITEST 13

For each of the following problems, decide which is the BEST of the choices given. If the exact numerical value is not one of the choices, select the choice that best approximates this value. Then fill in the corresponding circle on the answer sheet.

Notes: (1) A scientific or graphing calculator will be necessary for answering some (but not all) of the questions in this test. For each question you will have to decide whether or not you should use a calculator.

(2) For some questions in this test you may have to decide whether your calculator should be in the radian mode or the degree mode.

(3) Figures that accompany problems in this test are intended to provide information useful in solving the problems. They are drawn as accurately as possible EXCEPT when it is stated in a specific problem that its figure is not drawn to scale. All figures lie in a plane unless otherwise indicated.

(4) Unless otherwise specified, the domain of any function $f$ is assumed to be the set of all real numbers $x$ for which $f(x)$ is a real number. The range of $f$ is assumed to be the set of all real numbers $f(x)$, where $x$ is in the domain of $f$.

(5) Reference information that may be useful in answering the questions in this test can be found on the page preceding Question 1.

USE THIS SPACE FOR SCRATCH WORK.

1. If the graph of quadratic function $y = ax^2 + bx + c$ is concave down, which of the following must be true?

(A) $b^2 - 4ac = 0$
(B) $c \neq 0$
(C) $b^2 - 4ac < 0$
(D) $b^2 - 4ac > 0$
(E) $a \neq 0$

**MATHEMATICS LEVEL 2 Test - *Continued***

USE THIS SPACE FOR SCRATCH WORK.

2. If $f(x)$ and $g(x)$ are polynomial functions defined for all real numbers and $f(x) = (x+2)g(x) + k$ for all $x$, then $k =$

(A) $f(-2)$
(B) $f(2)$
(C) $f(0)$
(D) 0
(E) None of the above

3. Which of the following functions satisfies $f(x) = f^{-1}(x)$?

I. $y = -x$

II. $y = \dfrac{1}{x-2} + 2$

III. $y = \ln(x)$

(A) I only
(B) II only
(C) I and II
(D) I and III
(E) I, II, and III

## MATHEMATICS LEVEL 2 Test - *Continued*

USE THIS SPACE FOR SCRATCH WORK.

4. If $f(x) = \cos(x)$, $g(x) = -\cos(x)$, and $f(h(x)) = g(x)$, then $h(x) =$

(A) $\frac{\pi}{2} - x$

(B) $\frac{\pi}{2} + x$

(C) $\pi - x$

(D) $2\pi - x$

(E) $2\pi + x$

5. If $|x+3| = 5$, then $|x-2|$ could be

(A) 6
(B) 8
(C) 10
(D) 12
(E) 14

6. Which of the following must be true for $f(x) = a^x$ where $a > 0$?

(A) $f(x) + f(y) = f(xy)$.

(B) $\frac{f(x)}{f(y)} = f(x-y)$.

(C) If $x < y$, then $f(x) < f(y)$.

(D) $f(x^m) = mf(x)$ for all $m$.

(E) The graph intersects the $x$-axis at least once.

**MATHEMATICS LEVEL 2 Test - *Continued***

USE THIS SPACE FOR SCRATCH WORK

7. Which of the following is the equation whose graph is the set of points equidistant from the points $A(-2,1,5)$ and $B(-2,4,5)$?

(A) $y=\frac{1}{2}$

(B) $y=\frac{3}{2}$

(C) $x=\frac{3}{2}$

(D) $x=\frac{2}{3}$

(E) $y=\frac{5}{2}$

8. In space, three distinct points $A$, $B$, and $C$ are collinear, in that order. If $A(1,2,0)$, $C(-3,5,2)$ are given and $AB:BC=3:1$, what must be the coordinates of point $B$?

(A) $\left(-2,\frac{17}{4},\frac{3}{2}\right)$

(B) $\left(-4,\frac{17}{2},3\right)$

(C) $\left(-5,\frac{13}{2},3\right)$

(D) $\left(\frac{3}{2},2,\frac{17}{4}\right)$

(E) $\left(-2,\frac{13}{2},\frac{17}{2}\right)$

**MATHEMATICS LEVEL 2 Test - *Continued***

USE THIS SPACE FOR SCRATCH WORK

9. There are 9 distinct cookies baked in the oven. How many different ways are there to divide these cookies into three servings of four, three, and two cookies?

(A) 630
(B) 720
(C) 1260
(D) 3520
(E) 3024

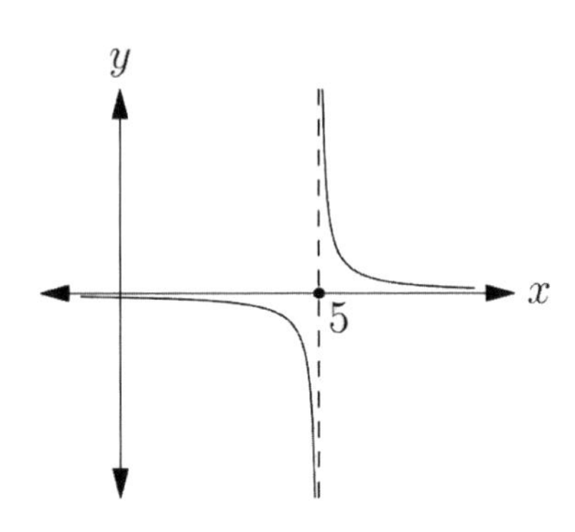


10. What value does $y = \dfrac{1}{2x-10}$ approach as $x$ approaches 5?

(A) 2
(B) $\dfrac{1}{10}$
(C) $-\dfrac{1}{10}$
(E) $-2$
(E) None of the above

**MATHEMATICS LEVEL 2 Test - *Continued***

USE THIS SPACE FOR SCRATCH WORK

11. What is the probability of getting three odd numbers when choosing three distinct numbers from $\{1,2,3,4,5,6,7,8,9,10,11,12,13\}$?

(A) 0.09
(B) 0.12
(C) 0.19
(D) 0.23
(E) 0.32

12. The difference between the maximum and the minimum value of $4\sin\left(\frac{x}{3}\right)\cos\left(\frac{x}{3}\right)$ is equal to

(A) 2 (B) $\frac{8}{3}$ (C) 4 (D) 8 (E) 16

13. If $\log(a) = 2\log(b)$, where $a$ and $b$ are integers between $-25$ and 25, not inclusive, what is the possible number of pairs $(a,b)$?

(A) 0 (B) 4 (C) 9 (D) 10 (E) 25

**MATHEMATICS LEVEL 2 Test - *Continued***

USE THIS SPACE FOR SCRATCH WORK

14. If the system of equations $\begin{cases} ax+2y=1 \\ bx-y=2 \end{cases}$ has a unique solution, which of the following CANNOT be $\begin{vmatrix} a & 2 \\ b & -1 \end{vmatrix}$?

(A) $-2$ (B) $-1$ (C) 0 (D) 1 (E) 2

15. Suppose an apartment is on fire. If the least distance between the fire engine and the building is 55 feet, for the safety procedure, what is the minimum length of the ladder if it is raised at $65°$ from the ground level?

(A) 112 feet
(B) 130 feet
(C) 242 feet
(D) 486 feet
(E) 728 feet

## Answerkey to Minitest 13

1. (E)

2. (A)

3. (C)

4. (C)

5. (C)

6. (B)

7. (E)

8. (A)

9. (C)

10. (E)

11. (B)

12. (C)

13. (B)

14. (C)

15. (B)

## Solution

1. The graph of $y = ax^2 + bx + c$ is concave down, so $a < 0$, which means $a \neq 0$.

2. Since $f(x) = (x+2)g(x) + k$, then $f(-2) = (-2+2)g(-2) + k = 0 + k = k$.

3. If $f(x) = f^{-1}(x)$, then $f(f(x)) = x$. Hence, I and II work, while III does not work.

4. $\cos(h(x)) = -\cos(x)$ if $\cos(\pi - x) = -\cos(x)$. Therefore, $\pi - x = h(x)$. This identity is validated because the graph of $y = \cos(x)$ is symmetric about the point $\left(\frac{\pi}{2}, 0\right)$.

5. Since $|x+3| = 5$, $x = 2, -8$. Therefore, $|x-2| = 0, 10$.

6. $y = a^x$ satisfies $f(a)f(b) = f(a+b)$. Hence, $f(x) = f(x-y)f(y)$ must be true.

7.

$$
\begin{aligned}
\sqrt{(x+2)^2 + (y-1)^2 + (z-5)^2} &= \sqrt{(x+2)^2 + (y-4)^2 + (z-5)^2} \\
(x+2)^2 + (y-1)^2 + (z-5)^2 &= (x+2)^2 + (y-4)^2 + (z-5)^2 \\
(y-1)^2 &= (y-4)^2 \\
y^2 - 2y + 1 &= y^2 - 8y + 16 \\
6y &= 15 \\
y &= \frac{15}{6} \\
y &= \frac{5}{2}
\end{aligned}
$$

8. Since $A$, $B$, and $C$ are collinear, where $AB : BC = 3 : 1$, if $b_x$ is the $x$-coordinate of $B$, then $b_x - 1 : -3 - b_x = 3 : 1$, so $-9 - 3b_x = b_x - 1$. Hence, $4b_x = -8$, so $b_x = -2$. Likewise, if $b_y$ is the $y$-coordinate of $B$, then $b_y - 2 : 5 - b_y = 3 : 1$, so $15 - 3b_y = b_y - 2$. Therefore, $4b_y = 17 \implies b_y = \frac{17}{4}$. Hence, the answer must be (A).

9. This is a typical partition question. Choose 4 items out of 9 items. Then, choose 3 items from the rest, i.e., $\binom{9}{4} \times \binom{5}{3} \times \binom{2}{2} = 1{,}260$.

10. $\lim_{x \to 5} \frac{1}{2x - 10}$ does not exist because the graph of $y = \frac{1}{2x - 10}$ has the vertical asymptote at $x = 5$.

11. $\frac{7}{13} \times \frac{6}{12} \times \frac{5}{10} = \frac{7}{52}$.

12. Remember that

$$\sin(2x) = 2\sin(x)\cos(x)$$

In other words, $2\sin(\frac{2x}{3}) = 4\sin(\frac{x}{3})\cos(\frac{x}{3})$, so the amplitude is 2. Since the difference between maximum and minimum is twice the amplitude, the answer is 4.

13. $\log(a) = 2\log(b) \implies \log(a) = \log(b^2)$. Hence, $a = b^2$ where $a, b > 0$. Thus, we can substitute values to find out that $(a,b) = (1,1), (4,2), (9,3)$, and $(16,4)$.

14. Since the system of linear equations has a unique solution, $a(-1) - 2(b) \neq 0$. Thus, the determinant cannot be 0.

15. Given a right triangle formed by the ladder, the ground, and the building, the angle is $65^\circ$ and the adjacent length is 55. Hence, if we call the ladder length $x$, then $\cos(65^\circ) = \frac{55}{x}$, so $x = \frac{55}{\cos(65^\circ)} \approx 130$.

# Minitest 14

**Minitest 14** includes

- Space Geometry
- Number Theory
- Logic
- Linear Functions and Equations
- Polar Coordinates
- Plane Geometry
- Trigonometric Functions
- Probability
- Rational Functions and Equations
- Trigonometric Identities
- Trigonometric Equations

# MATHEMATICS LEVEL 2 MINITEST 14

REFERENCE INFORMATION

THE FOLLOWING INFORMATION IS FOR YOUR REFERERENCE IN ANSWERING SOME OF THE QUESTIONS IN THE TEST.

Volume of a right circular cone with radius $r$ and height $h$ : $V = \frac{1}{3}\pi r^2 h$

Volume of a sphere with radius $r$ : $V = \frac{4}{3}\pi r^3$

Volume of a pyramid with base area $B$ and height $h$ : $V = \frac{1}{3}Bh$

Surface Area of a sphere with radius $r$ : $S = 4\pi r^2$

DO NOT DETACH FROM THE BOOK.

## MATHEMATICS LEVEL 2 MINITEST 14

For each of the following problems, decide which is the BEST of the choices given. If the exact numerical value is not one of the choices, select the choice that best approximates this value. Then fill in the corresponding circle on the answer sheet.

Notes: (1) A scientific or graphing calculator will be necessary for answering some (but not all) of the questions in this test. For each question you will have to decide whether or not you should use a calculator.

(2) For some questions in this test you may have to decide whether your calculator should be in the radian mode or the degree mode.

(3) Figures that accompany problems in this test are intended to provide information useful in solving the problems. They are drawn as accurately as possible EXCEPT when it is stated in a specific problem that its figure is not drawn to scale. All figures lie in a plane unless otherwise indicated.

(4) Unless otherwise specified, the domain of any function $f$ is assumed to be the set of all real numbers $x$ for which $f(x)$ is a real number. The range of $f$ is assumed to be the set of all real numbers $f(x)$, where $x$ is in the domain of $f$.

(5) Reference information that may be useful in answering the questions in this test can be found on the page preceding Question 1.

1. If $r$ is doubled and $h$ is tripled, the volume of a new circular cylinder is how many times that of the original cylinder?

USE THIS SPACE FOR SCRATCH WORK.

(A) 2
(B) 3
(C) 4
(D) 6
(E) 12

**MATHEMATICS LEVEL 2 Test - *Continued***

USE THIS SPACE FOR SCRATCH WORK.

2. What is the sum of all positive integer values of $n$ such that the following expression

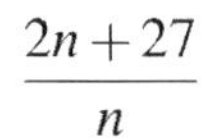

$$\frac{2n+27}{n}$$

is an integer?

(A) 1
(B) 4
(C) 13
(D) 36
(E) 40

3. Which of the following statements must be true?

I. If $x > 0$, then $x > 1$.

II. If $x \leq -1$, then $x \leq -2$.

III. If $x$ is a natural number, then $x$ is an integer.

(A) I only
(B) II only
(C) III only
(D) II and III
(E) I and III

**MATHEMATICS LEVEL 2 Test - *Continued***

USE THIS SPACE FOR SCRATCH WORK.

4. The graph of $y = f(x)$ is a straight line. If $f(3) \leq f(4)$, $f(7) \geq f(10)$ and $f(2) = 1$, what must be true?

(A) $f(0) = 0$
(B) $f(0) < 0$
(C) $f(1)f(2) < 0$
(D) $f(0) > 0$
(E) $f(0)$ is undefined.

5. There is a point $X$ in space, 5 inches directly above the plane $M$. To the nearest integer, what is the area, in square inches, of the portion $M$ that contains all points that are more than 10 inches and not more than 13 inches?

(A) 217
(B) 236
(C) 298
(D) 314
(E) 452

6. Which of the following function satisfies $f(x) = -|f(x)|$ for all $x$ in the domain?

(A) $y = x^2$
(B) $y = \ln(x)$
(C) $y = \frac{1}{x}$
(D) $y = \cos(x)$
(E) $y = -|x|$

**MATHEMATICS LEVEL 2 Test - *Continued***

USE THIS SPACE FOR SCRATCH WORK.

7. Which of the following could be the polar coordinates of the rectangular coordinates $\left(\frac{\sqrt{2}}{2}, \frac{\sqrt{2}}{2}\right)$?

(A) $\left(0, \frac{\pi}{4}\right)$
(B) $\left(1, -\frac{\pi}{4}\right)$
(C) $\left(-1, \frac{3\pi}{4}\right)$
(D) $\left(-1, -\frac{\pi}{4}\right)$
(E) $\left(1, \frac{9\pi}{4}\right)$

8. If $f(x) = 2x + 3$ and $g(x) = |x|$, which of the following CANNOT be in the range of $f(g(x))$?

(A) 6
(B) 5
(C) 4
(D) 3
(E) 2

9. How many non-congruent triangles are there if $m\angle B = 30^\circ$, $AB = 6$, and $AC = 4$?

(A) None
(B) One
(C) Two
(D) Three
(E) Infinitely many

**MATHEMATICS LEVEL 2 Test -** ***Continued***

USE THIS SPACE FOR SCRATCH WORK.

10. If the prime factorization of a natural number $n$ is given by $n = p^2q^3r$ for distinct primes $p$, $q$, and $r$, what is the total number of positive divisors of $n$?

(A) 6 (B) 8 (C) 12 (D) 24 (E) 36

11. What is the number of solutions for $\sin(2x) = \sin(x)$ from $x = 0$ to $x = 2\pi$, inclusive?

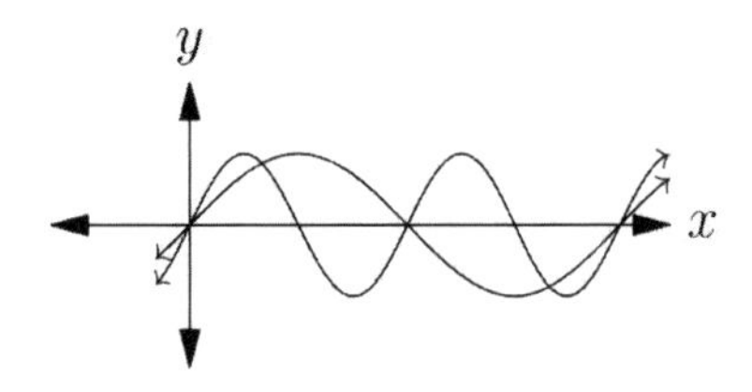


(A) 3 (B) 4 (C) 5 (D) 6 (E) 7

12. Timothy and Bob are smart. The probability of Timothy getting perfect score in SAT Math Level 2 is 0.91 and the probability of Bob getting the perfect score in SAT Math Level 2 is 0.94, respectively. If their probabilities are independent, what is the probability that at least one of them gets perfect score?

(A) 0.997
(B) 0.94
(C) 0.92
(D) 0.755
(E) 0.245

## MATHEMATICS LEVEL 2 Test - *Continued*

USE THIS SPACE FOR SCRATCH WORK.

13. If the difference of two numbers is 1 and the product of the two is 3, which of the following must be the positive difference of the reciprocals?

(A) 1 (B) 3 (C) 2 (D) $\frac{1}{3}$ (E) 4

14. If $\cos^2(\alpha)+\sin^2(\beta)=1.75$, then $\sin^2(\alpha)+\cos^2(\beta)=$

(A) 0
(B) 0.25
(C) 0.5
(D) 0.75
(E) 1

15. If $t$ is the number of years passed since September 2018, when does the population model for dolphins nearby *Nowhere* island, $P(t)=300-\sin\left(\frac{\pi}{2}t\right)$ reach its maximum for the first time?

(A) September 2019
(B) September 2020
(C) September 2021
(D) September 2022
(E) September 2023

## Answerkey to Minitest 14

1. (E)

2. (E)

3. (C)

4. (D)

5. (A)

6. (E)

7. (E)

8. (E)

9. (C)

10. (D)

11. (C)

12. (B)

13. (D)

14. (B)

15. (C)

## Solution

1. Instead of $r$, substitute $2r$. Likewise, instead of $h$, substitute $3h$. Then, the volume of the cylinder becomes $\pi(2r)^2(3h) = 12\pi r^2 h = 12V$. the new cylinder's volume is 12 times the original one's volume.

2. $\frac{2n+27}{n} = 2 + \frac{27}{n}$ is an integer if $n$ is a divisor of 27, i.e, $n = 1, 3, 9, 27$. Hence, $1+3+9+27 = 40$.

3. I is false because of $x = \frac{1}{2}$. II is false because of $x = -\frac{3}{2}$. III is true because there is no counterexample.

4. A linear function is either increasing, decreasing or constant. This function must be a constant function, so $f(x) = 1$. Hence, $f(0) > 0$.

5. Application of Pythagorean theorem implies that the smaller circle on the plane has the radius of $5\sqrt{3}$, whereas the bigger circle on the plane has the radius of 12. Hence, the area between the two concentric circles is $(12)^2\pi - (5\sqrt{3})^2\pi = 144\pi - 75\pi = 69\pi \approx 217$.

6. The condition $f(x) = -|f(x)|$ means that the graph of $y = f(x)$ is completely below the $x$-axis. The only function whose graph is below (or touching) the $x$-axis is $y = -|x|$.

7. $\left(1, \frac{9\pi}{4}\right) = \left(1\cos(\frac{9\pi}{4}, 1\sin(\frac{9\pi}{4}\right) = \left(\frac{\sqrt{2}}{2}, \frac{\sqrt{2}}{2}\right)$.

8. $f(g(x)) = 2|x| + 3 \geq 3$, so $y \neq 2$.

9. Remember SSA ambiguous cases. If $AC > BA\sin(30°)$, then there are two non-congruent triangles formed.

10. The total number of positive divisors of $n$ is $(2+1)(3+1)(1+1) = 24$.

11. $\sin(2x) = \sin(x)$ if $2x + x = 0, \pi, 3\pi, 5\pi$ and $2x - x = 0, 2\pi$. There are five points of intersections between the graph of $y = \sin(2x)$ and $y = \sin(x)$.

12. Let's use complementary counting, i.e., $1 - (0.09)(0.06) = 0.997$.

13. Let $x - y = 1$, and $xy = 3$. Then, $\frac{1}{y} - \frac{1}{x} = \frac{x-y}{xy} = \frac{1}{3}$.

14. Since $\cos^2(\alpha) + \sin^2(\alpha) + \cos^2(\beta) + \sin^2(\beta) = 2 = 1.75 + x$, where $x = \sin^2(\alpha) + \cos^2(\beta)$, then $x = 0.25$.

15. The population model reaches its maximum when $\sin\left(\frac{\pi}{2}t\right) = -1$, so $\frac{\pi}{2}t = \frac{3\pi}{2}, \frac{7\pi}{2}, \frac{11\pi}{2}, \cdots$. Hence, $t = 3$ is the first time when the population for dolphins reaches the maximum value, i.e, September 2021.

# Minitest 15

**Minitest 15** includes

- Polynomial Functions and Equations
- Plane Geometry
- Linear Functions and Equations
- Statistics
- Trigonometric Ratio
- Rational Functions and Equations
- Logarithm
- Logic
- Conic Sections
- Space Geometry

# MATHEMATICS LEVEL 2 MINITEST 15

REFERENCE INFORMATION

THE FOLLOWING INFORMATION IS FOR YOUR REFERERENCE IN ANSWERING SOME OF THE QUESTIONS IN THE TEST.

Volume of a right circular cone with radius $r$ and height $h$ : $V = \frac{1}{3}\pi r^2 h$

Volume of a sphere with radius $r$ : $V = \frac{4}{3}\pi r^3$

Volume of a pyramid with base area $B$ and height $h$ : $V = \frac{1}{3}Bh$

Surface Area of a sphere with radius $r$ : $S = 4\pi r^2$

DO NOT DETACH FROM THE BOOK.

## MATHEMATICS LEVEL 2 MINITEST 15

For each of the following problems, decide which is the BEST of the choices given. If the exact numerical value is not one of the choices, select the choice that best approximates this value. Then fill in the corresponding circle on the answer sheet.

Notes: (1) A scientific or graphing calculator will be necessary for answering some (but not all) of the questions in this test. For each question you will have to decide whether or not you should use a calculator.

(2) For some questions in this test you may have to decide whether your calculator should be in the radian mode or the degree mode.

(3) Figures that accompany problems in this test are intended to provide information useful in solving the problems. They are drawn as accurately as possible EXCEPT when it is stated in a specific problem that its figure is not drawn to scale. All figures lie in a plane unless otherwise indicated.

(4) Unless otherwise specified, the domain of any function $f$ is assumed to be the set of all real numbers $x$ for which $f(x)$ is a real number. The range of $f$ is assumed to be the set of all real numbers $f(x)$, where $x$ is in the domain of $f$.

(5) Reference information that may be useful in answering the questions in this test can be found on the page preceding Question 1.

1. If $t^3 - 8 = 4$, what is the value of $t$?

USE THIS SPACE FOR SCRATCH WORK.

(A) 1.45
(B) 2.29
(C) 3.43
(D) 4.00
(E) 5.78

**MATHEMATICS LEVEL 2 TEST — *Continued***

USE THIS SPACE FOR SCRATCH WORK.

2. In the $xy$-plane, which of the following is the set of points equidistant to two distinct intersecting lines?

(A) a circle
(B) an ellipse
(C) a single point
(D) a line
(E) two lines

3. If $f(x-2)=4x-2$, what is the value of $f(1)$?

(A) 3 (B) 5 (C) 10 (D) 12 (E) 15

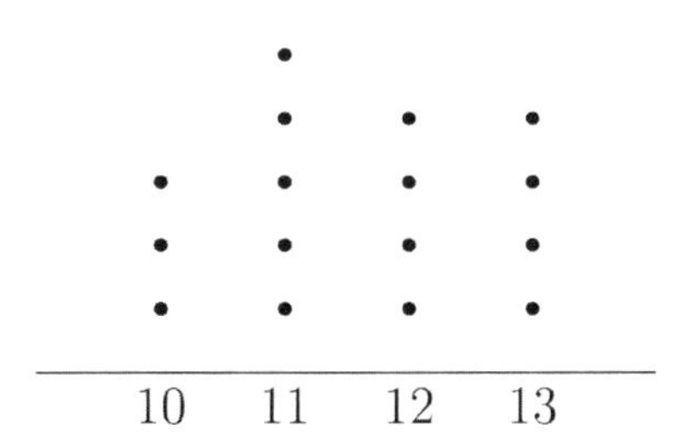


4. Given the dot plots above, what is the median?

(A) 10 (B) 11 (C) 11.5 (D) 12 (E) 13

**MATHEMATICS LEVEL 2 TEST — *Continued***

USE THIS SPACE FOR SCRATCH WORK.

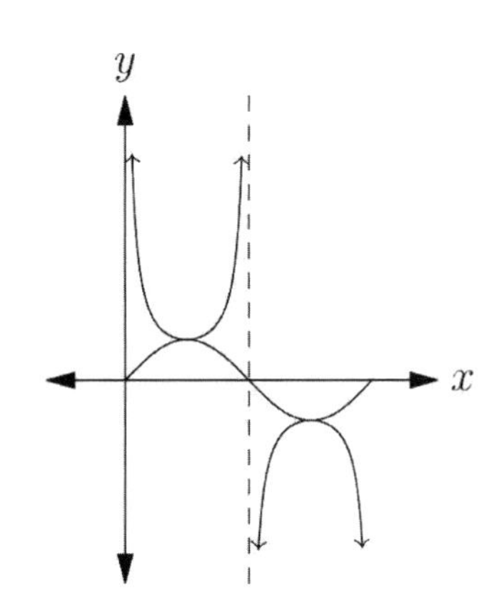


5. If $\sin(x) = 0.4312$, then $\csc(x) =$

(A) 0.9022
(B) 1.1249
(C) 2.3191
(D) 2.8312
(E) 2.9912

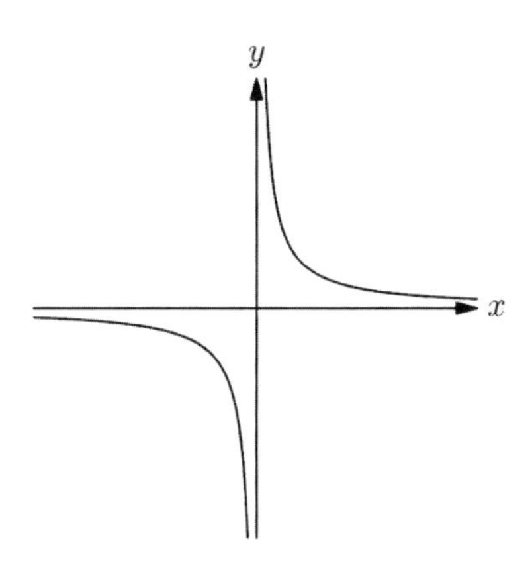


6. In the $xy$-plane, if $(4, 2)$ is on the graph of the hyperbola above $xy = k$, for some constant $k$, which of the following points must be on the graph?

(A) $(-4, 2)$
(B) $(4, -2)$
(C) $(2, 4)$
(D) $(-2, 4)$
(E) $(2, -4)$

**MATHEMATICS LEVEL 2 TEST — *Continued***

USE THIS SPACE FOR SCRATCH WORK.

7. What is the domain of the function $f$ given by $f(x) = \ln(|4 - x|)$?

(A) All real numbers
(B) All real numbers except 4
(C) All real numbers greater than 4
(D) All real numbers less than 4
(E) All real numbers less than or equal to 4

8. If $\dfrac{b+1}{b-1} = \dfrac{a-2}{a+2}$, then which of the following is equal to $a$?

(A) $2b$
(B) $b$
(C) $0$
(D) $-b$
(E) $-2b$

If $x > 3$, then $x \geq 5$.

9. Which of the following is the COUNTEREXAMPLE to the statement above?

(A) $x = 1$
(B) $x = 2$
(C) $x = 3$
(D) $x = 4$
(E) $x = 5$

**MATHEMATICS LEVEL 2 TEST — *Continued***

USE THIS SPACE FOR SCRATCH WORK.

10. Aaron who writes a book on SAT Math Level II earns a fixed amount of \$10,000 per month, plus \$5.99 each time the book is sold. If $n$ number of book is sold in January 2019, which of the following represents Aaron's earnings, in dollars, in that month?

(A) $(10{,}000+5.99)n$
(B) $5.99n$
(C) $10{,}000n+5.99$
(D) $10{,}000-5.99n$
(E) $10{,}000+5.99n$

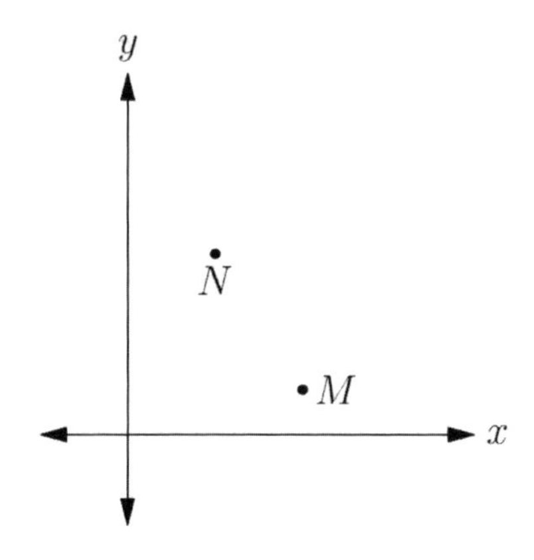


11. What is the slope of the line segment $\overline{NM}$ where $N=(2,4)$ and $M=(4,1)$?

(A) $\frac{3}{2}$
(B) $-\frac{2}{3}$
(C) $-\frac{3}{2}$
(D) $\frac{2}{3}$
(E) $\frac{3}{\sqrt{13}}$

**MATHEMATICS LEVEL 2 TEST — *Continued***

USE THIS SPACE FOR SCRATCH WORK.

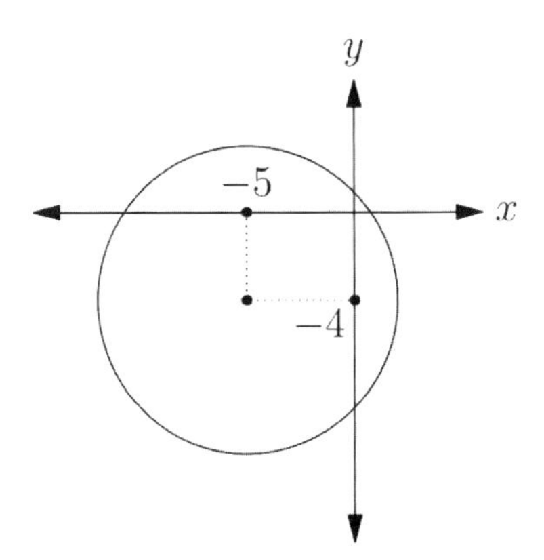


12. For which of the following values of $k$ does the equation $x^2+10x+y^2+8y=k$ represent a circle of radius 7?

(A) 1

(B) 8

(C) 7

(D) −8

(E) −1

13. A jar of coin contains exactly the same number of nickels, dimes and quarters. The total value of the coins in the jar is \$13.20. How many quarters and dimes are inside the jar?

(A) 15
(B) 25
(C) 33
(D) 40
(E) 66

**MATHEMATICS LEVEL 2 TEST — *Continued***

USE THIS SPACE FOR SCRATCH WORK.

14. Max and Julia planned going on a marathon run. Julia arrives late, so Max started running 16 minutes before Julia. He ran at an average rate of 9 minutes per mile, and Julia runs at an average rate of 8.25 minutes per mile. Assuming that the two started at the same location and ran the same route, how many hours and minutes will Julia take to catch up with Max?

(A) 1 hour and 47 minutes
(B) 2 hours and 13 minutes
(C) 2 hours and 38 minutes
(D) 2 hours and 56 minutes
(E) 3 hours and 11 minutes

15. In the $xyz$-coordinate system, which of the followings is NOT on the plane with equation $3x-2y+z=12$?

(A) $(0,0,12)$
(B) $(4,0,0)$
(C) $(1,-6,-3)$
(D) $(2,-1,4)$
(E) $(-1,6,0)$

## Answerkey to Minitest 15

1. (B)
2. (E)
3. (C)
4. (C)
5. (C)
6. (C)
7. (B)
8. (E)
9. (D)
10. (E)
11. (C)
12. (B)
13. (E)
14. (D)
15. (E)

## Solution

1. $t^3 = 12$, so $t = \sqrt[3]{12} \approx 2.29$.

2. The set of points equidistant to two intersecting lines is the two angle bisectors that are perpendicular to each other.

3. $f(x-2) = 4x-2 \implies f(1) = f(3-2) = 4(3)-2 = 10$.

4. The median is the average between 8th and 9th data value, so $\dfrac{11+12}{2} = 11.5$ is the median.

5. $\csc(x) = \dfrac{1}{\sin(x)} = \dfrac{1}{0.4312} \approx 2.3191$.

6. Since $xy = (4)(2) = 8$, Then, $(2,4)$ must be on the hyperbola.

7. $f(x) = \ln(|4-x|)$ has the domain value of $|4-x| > 0$. Hence, $x \neq 4$.

8. $1+\dfrac{2}{b-1} = 1-\dfrac{4}{a+2} \implies 2(a+2) = -4(b-1) \implies 2a+4 = -4b+4 \implies a = -2b$.

9. $x = 4$ is an obvious counterexample because $4 > 3$, but $4 < 5$.

10. $10{,}000 \text{ dollars} + \dfrac{5.99 \text{ dollars}}{1 \text{ book}} \times n \text{ books} =$ Aaron's earnings per month.

11. The slope of $\overline{NM}$ is $\dfrac{1-4}{4-2} = -\dfrac{3}{2}$.

12. $x^2+10x+25+y^2+8y+16 = k+25+16 = k+41 = 49$. Hence, $k = 8$.

13. $5n+10n+25n = 1320$, so $n = 33$. The number of quarters and dimes inside the jar must be $66(= 33+33)$ coins.

14. Let $t$ be the time Julia took to catch up with Max. Then,

$$(16+t) \text{ minutes } \times \frac{1 \text{ mile}}{9 \text{ minutes}} = t \text{ minutes } \times \frac{1 \text{ mile}}{8.25 \text{ minutes}}$$

where $t = 176$ (minutes), i.e. 2 hours and 56 minutes, is the time it took for Julia to catch up with him.

15.

- $3(0)-2(0)+12 = 12$.
- $3(4)-2(0)+0 = 12$.
- $3(1)-2(-6)-3 = 12$.
- $3(2)-2(-1)+4 = 12$.
- $3(-1)-2(6)+0 = -15 \neq 12$.

Therefore, $(-1,6,0)$ is not on the plane.

The Essential Workbook for SAT Math Level2

초판인쇄 2019년 10월 11일
초판발행 2019년 10월 11일

지은이 유하림
펴낸이 채종준
펴낸곳 한국학술정보(주)
주소 경기도 파주시 회동길 230(문발동)
전화 031) 908-3181(대표)
팩스 031) 908-3189
홈페이지 http://ebook.kstudy.com
전자우편 출판사업부 publish@kstudy.com
등록 제일산-115호(2000. 6. 19)

ISBN 978-89-268-9666-2 13410